THE ALPHA SEQUENCE

Electromagnetic Origin of the
Strong and Weak Nuclear Forces

THE ALPHA SEQUENCE

Electromagnetic Origin of the
Strong and Weak Nuclear Forces

MALCOLM H MAC GREGOR

Formerly of Lawrence Livermore National Laboratory, USA

World Scientific

NEW JERSEY · LONDON · SINGAPORE · BEIJING · SHANGHAI · HONG KONG · TAIPEI · CHENNAI · TOKYO

Published by

World Scientific Publishing Co. Pte. Ltd.

5 Toh Tuck Link, Singapore 596224

USA office: 27 Warren Street, Suite 401-402, Hackensack, NJ 07601

UK office: 57 Shelton Street, Covent Garden, London WC2H 9HE

Library of Congress Cataloging-in-Publication Data

Names: MacGregor, Malcolm H. (Malcolm Herbert), 1926–2019, author.
Title: The alpha sequence : electromagnetic origin of the strong and weak nuclear forces /
    Malcolm H. Mac Gregor.
Description: Hackensack, New Jersey : World Scientific, [2022] |
    Includes bibliographical references.
Identifiers: LCCN 2022009989 | ISBN 9789811252327 (hardback) |
    ISBN 9789811252334 (ebook for institutions) | ISBN 9789811252341 (ebook for individuals)
Subjects: LCSH: Electromagnetic interactions. | Fine-structure constant. |
    Particles (Nuclear physics) | Phenomenological theory (Physics)
Classification: LCC QC794.8.E4 M33 2022 | DDC 539.7/54--dc23/eng20220517
LC record available at https://lccn.loc.gov/2022009989

British Library Cataloguing-in-Publication Data
A catalogue record for this book is available from the British Library.

For any available supplementary material, please visit
https://www.worldscientific.com/worldscibooks/10.1142/12724#t=suppl

Typeset by Stallion Press
Email: enquiries@stallionpress.com

Für Elise

*The mystery about α is actually a double mystery. The first mystery —
the origin of its numerical value α ≈ 1/137 — has been recognized and
discussed for decades. The second mystery — the range of its domain — is
generally unrecognized.*

— Malcolm Mac Gregor,
Wikipedia, *Fine-structure constant*, Quotes:
The Power of Alpha (World Scientific, Singapore, 2007), p. 69.

*...particle lifetimes represent an ideal property to analyze, in the sense that
the mean lifetime t applies to all types of particle, so they can all be studied
together in one comprehensive analysis.*

— Malcolm Mac Gregor
The Alpha Sequence (World Scientific, Singapore, 2022), p. 5.

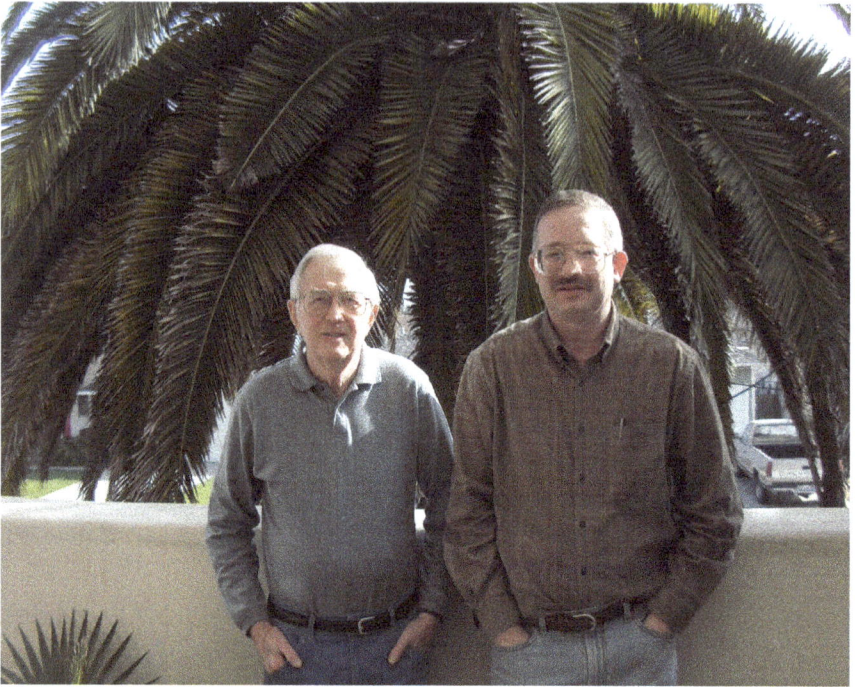

Malcolm H. Mac Gregor (*left*) and David A. Akers (*right*) in 2006 at the Mac Gregor's residence in Santa Cruz, California. (Photographer: Eleanor Mac Gregor)

Preface

With the passing of Dr. Malcolm H. Mac Gregor in 2019 Eleanor Mac Gregor contacted me with the request to review and to edit the manuscript before its publication.

Since I am very familiar with the scientific works of Malcolm Mac Gregor, as I had previously reviewed, edited, and made comments for his earlier book *The Power of Alpha*, I thought it would benefit the scientific community for me to undertake the task of editing *The Alpha Sequence*.

Conformity to orthodoxy has led many particle physicists to not look at the original idea that particle masses are linked to the mass of the electron and to the fine structure constant. Malcolm Mac Gregor was one of those who originally discovered this relationship. In fact, he had a personalized California vehicle registration plate with the number "70 MeV," as it is related to the fine structure constant.

It was a privilege to edit and to make comments on Malcolm's book The Power of Alpha in 2006. We met and corresponded throughout the years, even after his retirement from Lawrence Livermore National Laboratory, and we respected each other's ideas on elementary particle physics. It is, likewise, a privilege to review and to edit the manuscript for *The Alpha Sequence*.

David Akers
Lockheed Martin Skunk Works
Palmdale, California, 2019

Contents

Chapter 3.   The Experimental α-Quantization of Elementary Particle Energies and Masses, where $E = mc^2$ 41

List of Figures

List of Tables

Chapter 1　The Mysterious Fine Structure Constant $\alpha \sim 1/137$

One of the greatest mysteries in present-day physics is the role played by the fine structure constant

$$\alpha = \frac{e^2}{\hbar c} \approx \frac{1}{137.036}. \tag{1.1}$$

Properly speaking, α is not a fundamental constant, but rather a ratio of three fundamental constants, the electron charge e, Planck's constant h, and the speed of light c. A significant feature of α is that this ratio is dimensionless, so that it is the same in all systems of units.

The constant α plays a central role in quantum electrodynamics, QED, where it defines the coupling interaction strength between electrons and photons. It was first recognized as an important constant by Arnold Sommerfeld in 1916, who noted its occurrence in the velocity and binding energy of the electron in the lowest Bohr orbit of the hydrogen atom, and also in the fine-structure Zeeman splitting of atomic spectral lines in a magnetic field.

A puzzling feature of the constant α is its reciprocal numerical value $\alpha^{-1} \sim 137$. The precise value of $1/\alpha$, as obtained from the experimental values for $e, \hbar,$ and c is [1, 2]

$$137.035999139 \pm 31. \tag{1.2}$$

It was also measured directly in a 2008 precision experiment at Harvard, where an artificial atom was created by trapping a single electron in a positively-charged electrode, applying a calibrated magnetic field, and measuring the light produced as it jumped between two split levels. This experiment yielded the value [3, 4]

$$137.035999084. \tag{1.3}$$

Hence, the commonly-quoted value $1/\alpha \approx 137.036$ has an accuracy of almost nine significant figures.

Why the number 137 should appear in such an important constant has long defied explanation. Leon Lederman described this situation as follows [5]: "...this one number, 137, contains the crux of electromagnetism (the electron), relativity (the velocity of light), and quantum theory (Planck's constant). It would be less unsettling if the relationship between all of these important concepts turned out to be one or three or maybe a multiple of pi. But 137?". Lederman's home while he was the director of Fermilab was a 150-year-old farm house located at 137 Eola Road.

Richard Feynman also commented about the number 137 [6]: "It has been a mystery ever since it was discovered more than 50 years ago, and all good theoretical physicists put this number up on their wall and worry about it. ...It's one of the **greatest** damn mysteries of physics: a **magic number** that comes to us with no understanding by man. You might say the 'hand of God' wrote that number, and we don't know how He pushed His pen."

This discussion about the constant α involves a feature that has special pertinence for the investigations carried out in the present book, as described in a Wikipedia article on the fine structure constant:

> *"The mystery about α is actually a double mystery. The first mystery — the origin of its numerical value $\alpha \approx 1/137$ — has been recognized and discussed for decades. The second mystery — the range of its domain — is generally unrecognized."*
>
> Malcolm Mac Gregor,
> Wikipedia, *Fine-structure constant*, Quotes:
> *The Power of Alpha,*
> World Scientific, Singapore (2007), p. 69.

The present book is a phenomenological investigation of the experimental evidence, which reveals a clear-cut α-quantization of the elementary particle lifetimes, and also of the conjugate particle energies.

It is important to note here that the value $\alpha^{-1} \sim 137$ we have been discussing here for the fine structure coupling constant is the *renormalized* value that applies to *low-energy* processes. In particular, it is the value that constitutes the coupling between a real electron and a real photon in QED. There is also a *running* coupling constant $\alpha(q^2)$ that applies to the *high-energy* processes in quantum chromodynamics, QCD, which involve the interactions between quarks and gluons in the Standard Model of physics. The gluon is the carrier of the *strong force* that binds quarks together in a particle. It has the unique property that the binding energy gets stronger as quarks are pulled apart. This "rubber band" geometry confines the quarks, so that free quarks are not observed. The quarks themselves come in three *color charges*, which correspond in QCD to the role of the *electric charges* in QED. When interactions at high energies and high momentum transfers q^2 are studied, the effective length scale of the interaction *decreases*, and the color field forces themselves become logarithmically *stronger*. At the 80 GeV energy range of the W gauge boson, the momentum transfer is $q^2 \approx m_W^2$, and the running coupling constant has the value $\alpha(q^2) \approx 1/128$ [2].

The experimental data on elementary particles are summarized biennially in the *Review of Particle Properties* (RPP). The two most fundamental experimentally-determined numerical properties of a particle are its mean lifetime t and mass m. The 2018 RPP [1] lists 213 particles that have well-determined lifetimes and masses. The particle masses m are in turn proportional to the particle energies E, as defined by the Einstein equation $E = mc^2$. In the present analyses, we will employ particle energies rather than masses, since particle energies and lifetimes are conjugate quantities. The RPP experimental values for these particles are displayed here in Appendix A (Particle Mean Lifetimes) and Appendix B (Particle Energies Database).

Our main purpose in the present book is to investigate the role the fine structure constant α plays in the regularities exhibited by plots of the particle lifetimes and energies. The *strongly-interacting* particles, denoted collectively as *hadrons*, are the half-integer-spin baryons and integer-spin mesons (and their corresponding antiparticles), which are formed as combinations of the u, d, s, c, b Standard Model quarks (and their matching antiquarks). The lowest-mass state for each combination of quarks is the particle *ground state*, and higher-mass states with the same set of quark quantum numbers are short-lived *excited states* that mainly decay back to the ground states. The energy spectrum of these RPP particle states terminates at about 12 GeV. The experimental data contain roughly 40 hadrons that are identified as ground-state configurations. These are the states that we primarily focus on.

In addition to the *strongly-interacting hadronic* ground states, there are three *weakly-interacting leptons* — the electron, muon and tauon. There are also three *very-high-energy* states — the W and Z gauge bosons and top quark t, with energies above 80 GeV, whose lifetimes and energies have been accurately determined. It is of importance to determine if factor-of-137 regularities detected in the hadronic ground-states data can be extended to include the leptonic states and the very-high-energy particle states.

In carrying out a comprehensive data analysis, particle lifetimes are the most suitable property to employ, since they are specified by a single numerical quantity, the mean lifetime t in seconds, which applies to all types of particle. The RPP particle lifetime and energy data listed in Appendices A and B represent an essentially complete mapping of all the possible Standard Model quark combinations. Thus, they are in essence the finalized Standard Model data that present-day and future theorists and phenomenologists are going to have to work with. Any further experiments in the energy range up to 12 GeV will not substantially alter these data sets, except for possible exotic quark combinations that are outside of the range of the Standard Model configurations.

The particle lifetimes are analyzed in Chapter 2 to ascertain their dependence on the renormalized fine structure constant $\alpha^{-1} \sim 137$. As we will see, the particle data reveal their own story, which is reinforced in Chapter 3 by results obtained from the conjugate particle energies.

References

[1] M. Tanabashi *et al.* (Particle Data Group) *Phys. Rev. D* **98**, 030001 (2018) (URL: http://pdg.lbl.gov).
[2] Ref. [1], Table 1.1.
[3] D. Hanneke, S. Fogwell, and G. Gabrielse, *Phys. Rev. Lett.* **100**, 1207801 (2008).

[4] A. I. Miller, *137, Jung, Pauli, and the Pursuit of a Scientific Obsession* (Norton, New York, 2009) p. 248.

[5] L. Lederman with D. Teresi, *The God Particle* (Delta Books, New York, 1993), pp. 28–29.

[6] R. P. Feynman, *QED: The Strange Theory of Light and Matter* (Princeton University Press. 1985). p. 129.

Chapter 2 The Experimental α-Quantization of Lepton, Quark and Particle Mean Lifetimes

2.1 Introduction

There are 213 elementary particles in the RPP data compilation that have well-determined mean lifetimes t, including the stable electron and proton. These lifetimes are displayed in Appendix A. The long-lived particles, those with mean lifetimes longer than 10^{-21} sec (1 zeptosecond), have path lengths in a detector that are long enough to be measured directly. For the short-lived particles ($t < 1$ zsec), the Uncertainty Principle relationship $t = \hbar/\Delta E$ is used to determine t in terms of the mass/energy resonance width ΔE, where *energy* and *time* are conjugate quantum mechanical variables.

As mentioned in Chapter 1, particle lifetimes represent an ideal property to analyze, in the sense that the mean lifetime t applies to all types of particle, so they can all be studied together in one comprehensive analysis.

The hadrons that contain the quark flavors u, d, s, c, b (up, down, strange, charm, bottom) occur in the particle energy region below 12 GeV. Then there is a well-explored energy gap from 12 to 80 GeV, where no particles have been observed. Above this gap, the W and Z gauge bosons, Higgs resonance, and top quark t appear, with energies of about 80, 91, 125, and

173 GeV, respectively. No higher-energy particle states have been discovered as yet, even though the available particle excitation-energy range at the CERN Large Hadron Collider (LHC) has been extended up to a few hundred GeV.

2.2   The Four Elementary Particle Lifetime Zones

The list of 213 particle mean lifetimes displayed in Appendix A starts with the stable electron and proton, and it continues with 211 measured finite lifetimes that span the range from the 15 minute free-neutron lifetime down to the $\sim 10^{-24}$ sec lifetimes of the very high energy W and Z gauge boson and top quark t states. It is informative to divide these lifetimes into four lifetime zones, each of which represents a different type of particle decay. These particle lifetimes are displayed in Fig. 2.1, where they are plotted as exponents to the base 10 on a logarithmic abscissa.

Zone 1 contains the lifetimes that are longer than 10^{-7} sec. It includes just the hadronic free neutron and leptonic muon. As we will demonstrate, these particles fit into and extend the overall global lifetime grid, which we define and characterize by the shorter-lived pseudoscalar (ps) meson lifetime group that is displayed in Zones 2 and 3.

The Zone 2 lifetimes range from 10^{-7} down to 10^{-14} sec. This zone contains the *unpaired-quark* ground states, which are the mesons and baryons that are formed as various combinations of the (u, d), s, c, b quark flavors. These ground-state particles can only decay by *flavor-breaking* (electroweak) transitions, which accounts for their long lifetimes. Their lifetimes fall into four discrete flavor groups that correspond to the (ps), s, b, c labels shown in Zone 2. Each group has an identifiable central lifetime (CL) that is determined by its dominant quark flavor. The (ps), s and b groups are

The experimental lifetime zones of 211 elementary particle states

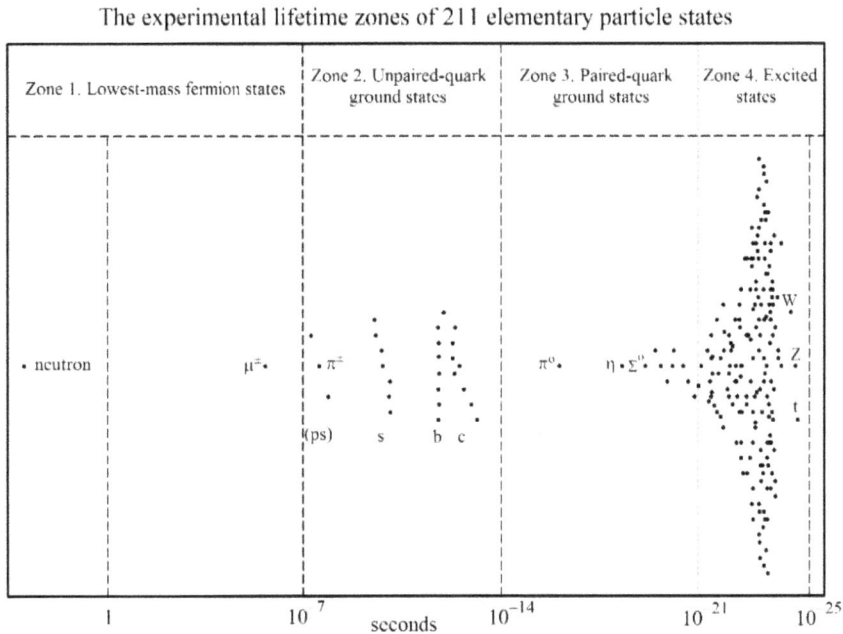

Figure 2.1 The 211 experimental particle lifetimes from Review of Particle Physics (2018) [1], plotted on a base 10 logarithmic lifetime grid that extends over 25 orders of magnitude. The four indicated lifetime zones each contain particles with a different type of decay mechanism. An important feature of this lifetime plot is that no *rogue* particles appear as exceptions to the systematics.

widely separated, but the *c*-group, which also includes the $\tau^{\pm}$ lepton, has a central lifetime CL that is a just a factor of 3 shorter than the *b*-group CL lifetime. The *c*-group has several lifetimes that deviate from its CL by factors of 2, whereas the *b*-group lifetimes are all tightly bunched along its CL.

The Zone 3 lifetimes range from 10^{-14} down to 10^{-21} sec, and contain *paired-quark* ground-state configurations These have *flavor-conserving* decays, and hence much shorter lifetimes, than the *unpaired-quark* ground states of Zone 2. Also included are ground-state quark configurations that have flavor-conserving *radiative* decays.

Zone 4 contains the lifetimes that are shorter than a *zeptosecond* (10^{-21} sec). The particles in this zone are excited states of the particles in Zones 2 and 3. The 1 zsec upper-limit lifetime boundary serves as the line of demarcation between the ground-state and excited-state particles. The only ultra-short lifetimes — comparable to a *yoctosecond* (10^{-24} sec) — are those of the ultra-high-energy W and Z gauge bosons and top quark t particle states, which are each a few ysec.

The (ps), s, b, c separated flavor groups of *unpaired-quark ground states* displayed in Zone 2 of Fig. 2.1 are the most informative particle lifetimes to consider from the viewpoint of looking for a possible α dependence. In Sec. 2.3 we examine the (ps) pseudoscalar pion and kaon mesons, which are the lowest-mass and longest-lived hadronic states, and therefore probably the most basic, and were among the earliest particle states to be discovered. The lifetimes of these (ps) states are plotted in Figs. 2.2a, 2.2b, and 2.2c as exponents on a logarithmic abscissa to the base α. As it turns out, their lifetime properties serve to define an α-quantized global lifetime grid that applies to the spectrum of unstable elementary particles shown in Appendix A.

2.3   The α-Quantized Pseudoscalar Meson Lifetimes

In order to quantitatively determine the lifetime intervals displayed in Fig. 2.1, we select the long-lived pseudoscalar meson lifetimes and plot them here in detail, using an α-quantized logarithmic abscissa for the lifetime grid. The *non-strange* $\pi^{\pm}, \pi^{0}, \eta, \eta'$ mesons are shown in Fig. 2.2a, and the *strange* $K^{\pm}, K^{0}_{L}, K^{0}_{S}$ kaons are shown in Fig. 2.2b. The lifetimes are combined together in Fig. 2.2c. These are the lowest-mass and longest-lived hadronic particle states, and thus the most important ones to study. The data fits to these pseudoscalar mesons are sufficient to establish the

The α-spacings of the nonstrange pseudoscalar meson lifetimes

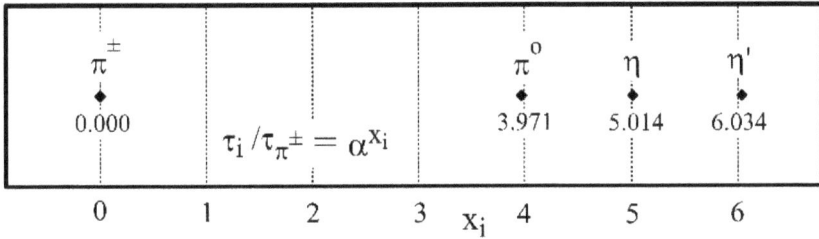

Figure 2.2a The nonstrange pionic lifetimes τ_i, plotted as ratios to the $\tau_{\pi\pm}$ reference lifetime and displayed on a logarithmic α-spaced lifetime grid. They illustrate an α^4 decrease in going from the flavor-breaking $\pi^\pm$ decay to the flavor-conserving π^0 radiative decay, and additional decreases by factors of α in going from the π^0 decay to the η and η' excited-state decays. The experimental value of the logarithm x_i to the base α for each particle i is shown under the particle.

Factors of 2 and α in the strange pseudoscalar kaon lifetimes

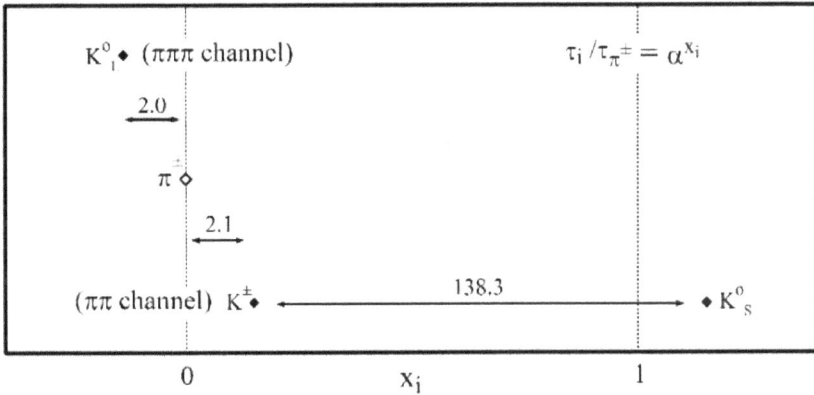

Figure 2.2b The strange kaon lifetimes, plotted as ratios to the $\pi^\pm$ lifetime. They show accurate factor-of-2 lifetime ratios between $K_L^0/\pi^\pm$ and $\pi^\pm/K^\pm$ pairs, and also a factor of α decrease in going from the $K^\pm$ to the K_S^0 lifetime. The $K^\pm$ and K_S^0 decays are into $\pi\pi$, whereas the K_L^0 decay is into $\pi\pi\pi$.

The α-quantized pseudoscalar meson and kaon lifetimes

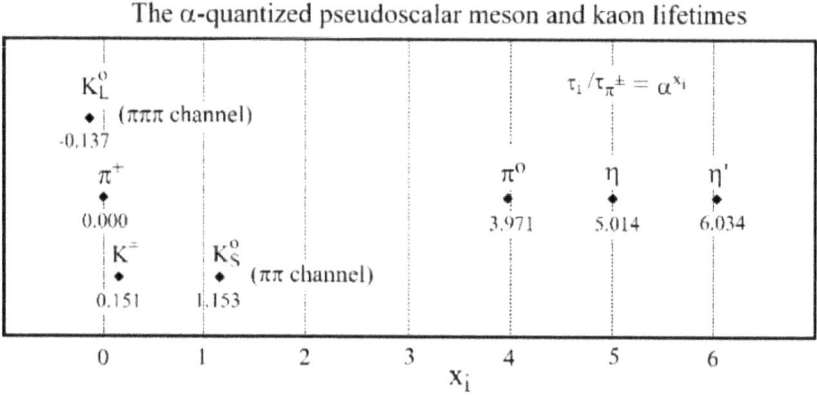

Figure 2.2c   The pion and kaon lifetimes, shown plotted together on the α-spaced lifetime grid. This α-grid accurately applies to the subsequently-discovered b quarks at $x_i = 2$, and, with an additional factor of 3 towards shorter lifetimes, also to the c quarks, as displayed in Fig. 2.4 below.

properties of an α-quantized logarithmic lifetime grid that applies to the entire spectrum of elementary particle ground states.

The lifetime systematics displayed in these three figures includes the following significant features, which carry over and also apply to the subsequently-discovered b-quark and c-quark particles:

(a) the selection of the $\pi^{\pm}$ meson as the reference lifetime, which is the central lifetime (CL) of the $K_L^0, \pi^{\pm}, K^{\pm}$ pseudoscalar (PS) meson triad;

(b) the fact that the values of the logarithms x_i to the base α shown in Fig. 2.2a for the π^0, η, η' meson triad are very close to integers, which indicates that an α-spaced lifetime grid anchored on the $\pi^{\pm}$ lifetime gives an accurate fit to the experimental data over a range of 6 powers of α, or 13 orders of magnitude;

(c) the factor-of-2 displacements of the K_L^0 and $K^{\pm}$ lifetimes from the CL, which are also observed in other particle states (Figs. 2.4–2.6);

(d) the factor of 138.3 lifetime ratio between the $K^{\pm}$ and K^0_S lifetimes, which share a common $\pi\pi$ decay channel, and which are each displaced from a CL line by a factor of 2;

(e) the α^4 lifetime ratio between the $\pi^{\pm}$ and π^0 lifetimes, which are flavor-breaking and flavor-conserving decays, respectively, and is a ratio that also occurs in the s, c and b lifetime groups;

(f) the fact that no lifetimes occur between the $\pi^{\pm}$ and π^0 lifetimes, so that this α^4 gap stands as a "lifetime desert region" which separates the flavor-conserving from the flavor-breaking decays in all of the lifetime groupings (Fig. 2.7).

Historically, the pseudoscalar mesons and s-quark hyperons were the first particles whose lifetimes revealed their α-quantized structure. The fact that the general (u, d, s)-quark lifetime features listed here are also evident in the higher-mass c-quark and b-quark lifetimes when plotted on the same α-spaced logarithmic lifetime grid lends credence to this comprehensive lifetime α-quantization grid. This observed lifetime α-quantization indicates that the conjugate α-quantized energy packets which create these particles in the first place (Chapter 3) also play a role in their decay rates — that is, in their intrinsic stability.

2.4   The Pseudoscalar Meson $\chi^2(S)$ Minimization Curve for the Lifetime Scaling Factor S

The lifetime regularities displayed by the $\pi^{\pm}$, π^0, η, η' nonstrange pseudoscalar mesons in Fig. 2.2a are sufficient to define an α-quantized lifetime grid that applies to all of the basic long-lived ground-state particles. It is of interest to illustrate the large span of lifetimes encompassed by these mesons. Expressing their lifetimes t in zeptoseconds (10^{-21} sec) gives the following table:

Table 2.1   Meson lifetimes in zeptoseconds.

$\pi^{\pm} = 26{,}033{,}000{,}000{,}000$	zsec
$\pi^{0} = 85{,}200$	zsec
$\eta = 502$	zsec
$\eta' = 3.32$	zsec

When a consistent lifetime formalism relates particles with such widely different time frames, it indicates that these quantities share a fundamental lifetime scaling factor S. We can obtain an experimental value for S mathematically by fitting it to the lifetime data, as we now describe.

It seems visually apparent from the spacings of these nonstrange pseudoscalar mesons lifetimes on the lifetime α-grid of Fig. 2.2a that they have a functional global dependence on the fine structure constant $\alpha \cong 1/137$. We can move beyond this qualitative display and obtain a quantitative evaluation of the accuracy of the lifetime scaling. This is accomplished by writing down an equation for the lifetime of each particle as a function of a variable scaling parameter $S \sim 137$, and then doing a series of $\chi^2(S)$ least-squares fits of these calculated lifetimes to the experimentally measured lifetimes as weighted by the uncertainties in these measurements [2]. In detail, we write down equations for the lifetimes $t_i(S)$ as ratios to the $\pi^{\pm}$ lifetime, compare the calculated values $t_i(S)$ to the experimental lifetimes $t_i(\mathrm{exper})$, and then find the value of S that minimizes the chi-squared sum

$$\chi^2(S) = \sum_i \left(\frac{t_i(S) - t_i(\mathrm{exper})}{\Delta t_i(\mathrm{exper})} \right)^2, \tag{2.1}$$

where $\Delta t_i(\mathrm{exper})$ is the experimental error of the ith measurement.

This method only works well in a statistical sense if all of the data points have roughly comparable error limits, which is the case here. Otherwise, the χ^2 sum is dominated by the fits to the data points that have very small error bars. The χ^2 analysis is carried out as follows. Using the $\pi^{\pm}$ meson as the reference lifetime, we write the π^0, η and η' lifetimes in the form of the equation $t_i/t_{\pi^{\pm}} = S^{-x_i}$, $x_i = 4, 5, 6$, respectively, and insert these values into Eq. (2.1). Then we vary S in order obtain the value that minimizes the χ^2 sum. The results are displayed in Fig. 2.3. As can be seen, the $\chi^2(S)$ minimum occurs at $S_{\min} = 139.1$, which closely agrees with the α^{-1} value 137.0. This demonstrates that it is the *renormalized* coupling constant $\alpha^{-1} \cong 137$, and not the smaller value of $S_{\min} = 128$, which we might expect to find if we were using the running coupling constant $\alpha(q^2)$ [3], that applies to these pseudoscalar lifetimes.

As we displayed in Table 2.1, the lifetime leap from the $\pi^{\pm}$ to the π^0 meson is more than eight orders of magnitude. To have this leap be almost precisely equal to α^4, and then to have it followed by an additional leap of α for each of the two higher excited states, is direct experimental evidence for the relevance of α in this lifetime domain. It should be kept in mind that the accurate *lifetime* 1/137 scaling factors shown in Fig. 2.2a

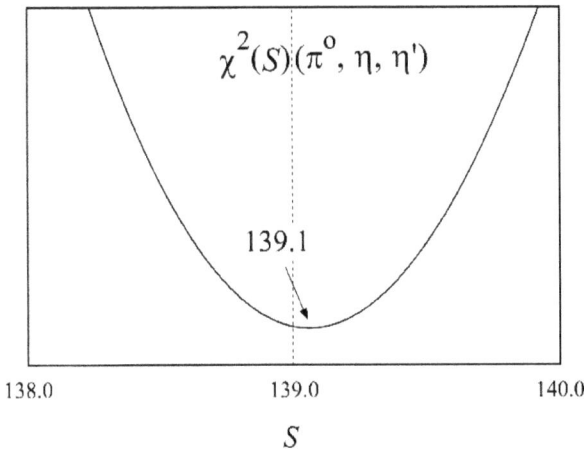

Figure 2.3 The experimental determination of the scaling factor S from $\chi^2(S)$ fits to the pseudoscalar π^+, π^0, η, η' meson lifetimes displayed in Fig. 2.2a.

for the pseudoscalar $\pi^{\pm}, \pi^0, \eta, \eta'$ mesons logically correlate with the accurate conjugate *mass/energy* scaling factors of 137 displayed in Table 3.2 of Chapter 3 for these same pseudoscalar particles. In the next section, we extend this lifetime systematics to include the c-quark and b-quark lifetime groups that were identified in Zone 2 of Fig. 2.1.

2.5   The Lifetime Quark Dominance Rule $c > b > s > (u, d)$

The long-lived hadronic *particle ground states* have the lifetimes that lie in Zone 2 of Fig. 2.1. As can be seen, they occur in four separated lifetime groups, which are labeled sequentially in Fig. 2.1 by the quark flavors (u, d), s, b and c. To illustrate these results in detail, we display the Zone 2 lifetimes in Fig. 2.4, where the particle states themselves are identified, and are plotted on an α-spaced logarithmic lifetime grid centered on the $\pi^{\pm}$ lifetime. As can be seen, the lifetimes are bunched into well-separated

α-quantized unpaired-quark lifetimes, showing c > b > s > (ud) quark dominance

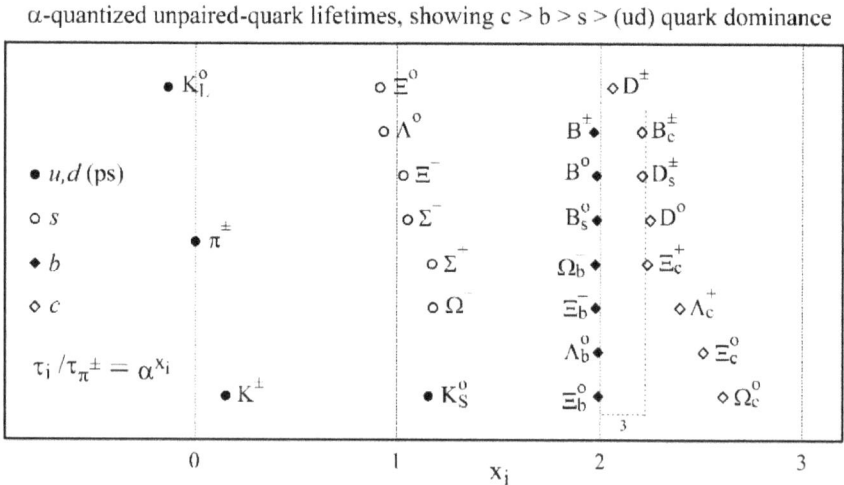

Figure 2.4   The Zone 2 particles of Fig. 2.1, which have lifetimes between 10^{-7} and 10^{-14} seconds. These lifetimes fall into four separate quark groupings: (u, d), s, b, c, as indicated by the lifetime symbols labeled at the left.

groups, each of which features a characteristic dominant quark sub-state. The description of these separated quark lifetime groups is as follows:

The (u, d) *group* consists of the $\pi^{\pm}, K^{\pm}, K_L^0, K_S^0$ pseudoscalar (ps) mesons, with the $\pi^{\pm}, K^{\pm}, K_L^0$ mesons centered on the $x_i = 0$ grid line (the *central lifetime* CL of the group), and the K_S^0 meson displaced by one power of α relative to the $K^{\pm}$ meson (see Fig. 2.2b).

The s *group* consists of the Λ, Σ, Ξ and Ω hyperons and K_S^0 meson, which are grouped along the $x_i = 1$ CL grid line.

The b *group* contains three b meson and four b baryon states, which are all tightly-bunched along the $x_i = 2$ CL grid line.

The c *group* contains four meson and four baryon states, four of which are grouped along a CL grid line that is displaced by a factor of 3 to the right of the $x_i = 2$ CL grid line.

It is noteworthy that no *maverick* particle lifetimes (lifetimes that lie outside of these bunched flavor groups) are observed in Zone 2.

These experimental lifetime groupings lead to several important empirical conclusions. The first is:

> *The lifetimes of the ground-state particles are determined by the intrinsic lifetimes of their quark substates.* (2.2)

Each quark flavor has its own characteristic lifetime. The particular quark in a particle that determines the particle lifetime is the one with a well-defined feature, which is the second empirical conclusion:

> *The lifetime of a ground-state particle is dominated by the lifetime of its shortest-lived quark.* (2.3)

The ordering of the relative magnitudes of the quark lifetimes is revealed by the ordering of the lifetime groups shown in Fig. 2.1, which leads to the third empirical conclusion:

$$\textit{The quark dominance rule is } c > b > s > (u, d) \tag{2.4}$$

The operation of the $c > b > s > (u, d)$ quark dominance rule is clearly revealed by the b-quark lifetimes shown in Fig. 2.4, where the following quark flavors occur:

$$\mathrm{B}^+ = bu; \quad \mathrm{B}^0 = bd; \quad \mathrm{B_S} = bs, \quad \mathrm{B_c} = bc;$$

$$\Lambda_\mathrm{b}^- = bdd; \quad \Xi_\mathrm{b}^0 = bsu; \quad \Xi_\mathrm{b}^- = bsd; \quad \Omega_\mathrm{b}^- = bss.$$

The two key lifetimes here are those of the $\mathrm{B_S} = bs$ and $\mathrm{B_c} = bc$ mesons. The $\mathrm{B_S}$ meson lifetime is firmly in the b-quark grouping, and the $\mathrm{B_c}$ meson lifetime is firmly in the c-quark grouping. Thus, the shortest lifetime in each case is the one that triggers the particle decay.

The one exception to the quark-dominance rule occurs in the K mesons lifetimes, and it provides some interesting information. As formulated in the pseudoscalar meson octet, the two neutral kaon states are identified as the particle-antiparticle pair $(\mathrm{K}^0, \bar{\mathrm{K}}^0)$ which carry the strangeness quantum numbers $s = -1$ and $s = +1$, respectively. However, these are not the states that are observed experimentally. Instead, the states that are seen are the $\mathrm{K_L^0} \to \pi\pi\pi$ and $\mathrm{K_S^0} \to \pi\pi$ mesons, which are theoretically treated as linear combinations of the K^0 and $\bar{\mathrm{K}}^0$ mesons. This implies that the $\mathrm{K_L^0}$ and $\mathrm{K_S^0}$ mesons each have mixed particle-antiparticle symmetry and mixed strangeness s. The lifetime systematics displayed in Fig. 2.4 provides three significant facts with respect to these neutral kaons:

(1) The K_L^0 meson fits in naturally as a member of the $K_L^0, \pi^\pm, K^\pm$ triplet in the (u, d) lifetime group on the $x_i = 0$ α-grid, and the K_S^0 meson matches the Σ^+ and Ω^- lifetimes on the $x_i = 1$ α-grid.

(2) The $K^\pm$ and K_S^0 mesons, which both have $\pi\pi$ decay modes, and whose lifetime ratio is 138.3, are each offset from the x_i grid lines by the same factor of 2, as displayed in Fig. 2.2b.

Thus, the K_S^0 meson has the short lifetime that we associate with the *strange* quark s group of lifetimes, and the K_L^0 and $K^\pm$ mesons have the factor-of-137 longer lifetimes that we associate with the *non-strange* (u, d) quark pair group of lifetimes. These facts have a bearing on the particle–antiparticle mixing and the strangeness quantum number mixing that seems inherent in the (K_L^0, K_S^0) neutral kaon doublet.

Perhaps the most important point to emphasize here is something that does *not* occur:

(3) The K_L^0 and K_S^0 kaons are regarded as having mixed combinations of strangeness and mixed particle–antiparticle symmetry, and yet they do *not* exhibit *rogue* lifetimes that do not fit in with the well-separated lifetimes of the other Zone 2 particles shown in Fig. 2.4, which all have well-defined quark flavor and baryon quantum numbers. On the contrary, they fit into and help define the α-spaced lifetime grid that is anchored on the $\pi^\pm$ lifetime. These empirical results should provide interesting material for particle theorists to explore.

It seems appropriate for me to inject a short historical note here. In the mid-1950's, I was visiting the Lawrence Berkeley Radiation Laboratory on the day when Murray Gell-Mann arrived there from Cal Tech and delivered a seminar on his newly-postulated strangeness quantum number. I can still picture him writing the letter S on the blackboard. He explained how the conservation of "strangeness" could account for the puzzling fact that K^-K^0 pairs (so-called "associated production" pairs) have large production cross sections, which indicated a "strong-interaction" process, but also have very long lifetimes, which are characteristic of "weak" decays. The explanation is that the strong *associated-production* interactions conserve strangeness, whereas the separate K^- and K^0 decays do not conserve strangeness [4]. Breaking strangeness can only be accomplished via the weak interaction mechanism, which greatly slows down the decay process. The introduction of the strangeness quantum number S preceded the discovery of the quark model. When, a few years later, the (u, d, s) quark model was set forth, the strangeness quantum number was subsequently (but not immediately) identified as being carried by the s quark.

2.6   Quark Group Central Lifetimes (CL) and Factor of 2-3-4 Deviations from the CL

We now move on to the more detailed features of the Zone 2 lifetime flavor groups shown in Fig. 2.4. These are displayed in Fig. 2.5.

The Zone 2 lifetimes of Fig. 2.1, which feature weak-interaction unpaired-quark ground-state decays, were plotted in detail in Fig. 2.4 in order to demonstrate the workings of the $c > b > s > (u, d)$ quark rule for the intrinsic quark lifetimes. These lifetimes are replotted here in Fig. 2.5, where the *central-lifetime* (CL) particles are singled out, and where the important fact is illustrated that the particles which do *not* fall on the CL are separated from it by *near-integer* lifetime factors of approximately 2, 3, or 4. The (ps), (s), (b) and (c) central lifetimes in Fig. 2.5 have the following features:

α-quantized flavor-group central lifetimes (CL) and factor of 2-3-4 deviations

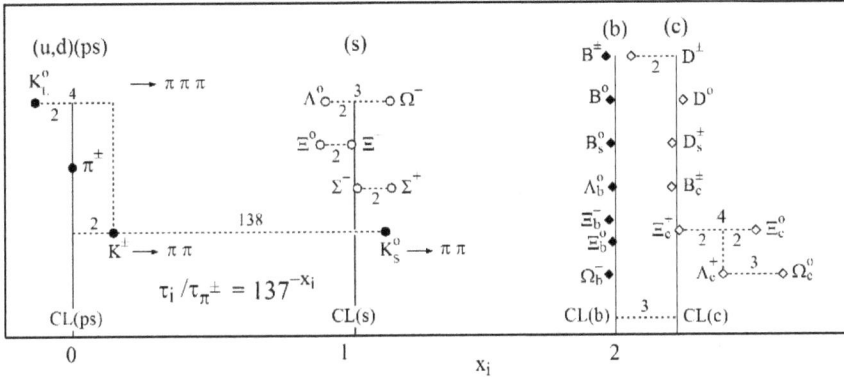

Figure 2.5 The Zone 2 lifetimes of Fig. 2.4 plotted here on the same logarithmic α-grid abscissa. The four clearly-separated flavor groups still appear, but are internally rearranged to bring out the group substructures. Vertical solid lines delineate the *central lifetime* (CL) in each quark flavor group. Dashed lines illustrate the factor of 2, 3 and 4 deviations of the other particle lifetimes from the flavor group CL. These deviations are shown arrayed together in Fig. 2.6.

The (u, d)-quark pseudoscalar (ps) meson vertical CL(ps) line is at the $x_i = 0$ logarithm of the $\pi^{\pm}$ pseudoscalar meson, which is the reference lifetime that anchors the α-spaced lifetime grid.

The s-quark central lifetime CL(s), which is taken to be the average of the Σ^- and Ξ^- lifetimes, is slightly shorter than the lifetime that corresponds to the $x_i = 1$ lifetime logarithm.

The b-quark central lifetime CL(b) does not continue this s-quark deviation; instead, it closely matches the $x_i = 2$ lifetime grid. Also, in contrast to the other quark lifetime groups, the b-quark particles (except the B_c) all have lifetimes that match the CL(b) lifetime. In detail, the seven b-quark CL(b) particles shown in Fig. 2.5 have an average α-grid logarithm value of $x_i = 1.98$.

The c-quark lifetimes are the only ones that do not fit on the α-quantized lifetime grid: they are displaced toward shorter lifetimes by a factor of 3 with respect to the b-quark lifetimes. Experimentally, the $(D^{\circ}, D_s, B_c, \Xi_c^-)$ CL(c) group of particles have an average α-grid value of $x_i = 2.2229$, which corresponds to a numerical displacement by a factor of 2.994 from the $x_i = 2$ logarithm. Thus, a systematic b-quark to c-quark shift toward shorter lifetimes by a factor of almost exactly 3 is indicated by the experimental lifetime data.

The lifetimes in Fig. 2.5 that do not fall on the central lifetimes exhibit a lifetime substructure in which the lifetime ratios between related particles occur with approximately integer values of 2, 3 or 4. Figure 2.5 contains eight factor-of-2 ratios, two factor-of-3 ratios, and two factor-of-4 ratios. The experimental values for these lifetime ratios are displayed graphically in Fig. 2.6, where they are grouped together and averaged. The eight factor-of-2 ratios have an *average displacement ratio* (ADR) of 1.95. The two ratios, $\Lambda^{\circ} - \Omega^-$ and $\Lambda_c^+ - \Omega_c^0$ have an ADR of 3.05. The $K_L^{\circ} - K^{\pm}$ and $\Xi_c^+ - \Xi_c^{\circ}$ matched pairs have an ADR of 4.04. Thus, individual lifetime deviations from the CL values can be up to 10%, but each ADR group value is within 2% of being exact. Also, there are no rogue lifetimes that do not fit into this systematics. Hence, these approximately-integer lifetime ratios seem significant.

From a phenomenological point of view, the *dominant quark flavor* determines the lifetime group in which a particular particle belongs. Within the group, there is a central lifetime CL that acts as the *centroid* for the distribution of lifetimes within the group. This centroid lifetime can be considered to be the *intrinsic flavor lifetime* of the quark. Numerical experimental values for these intrinsic quark flavor lifetimes are summarized in Table 2.2.

Factor of 2-3-4 deviations from the central lifetime CL

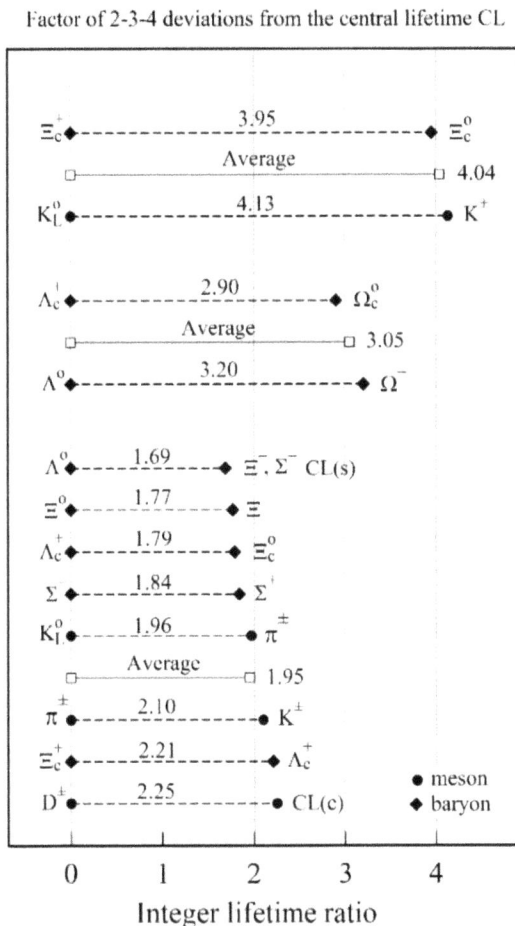

Integer lifetime ratio

Figure 2.6   The factor of 2 or 3 or 4 deviation ratios of the particle lifetimes from the central lifetimes that were displayed in Fig. 2.5 are plotted together here in separate 2, 3, 4 groups. The individual deviation ratios are not exact integers, but the average deviation ratio (ADR) within each group is close to an integer. This suggests that additional small factors also affect the lifetimes.

Particle deviations from the intrinsic quark lifetimes are by small approximately-integer factors, which can be positive or negative. This lifetime granularity may be due to a granularity in the internal structure of the quark, or to the available number of decay channels. The fact that the

Table 2.2   The experimental intrinsic quark flavor (CL) lifetimes.

$(u, d) = 2.60 \times 10^{-8}$ sec	$(\pi^{\pm}$ lifetime$)$
$s = 1.56 \times 10^{-10}$ sec	(Σ, Ξ^{-}) average lifetime
$b = 1.53 \times 10^{-12}$ sec	$(B^{+}, B^{\circ}, B_s, \Lambda_b^{-}, \Xi_b^{0}, \Xi_b^{-}, \Omega_b^{-})$ average
$c = 5.10 \times 10^{-13}$ sec	$(D^{\circ}, D_s, B_c, \Xi_c^{-})$ average lifetime

deviations are not exact integers suggests that other small effects also affect the decay process, such as the available phase space for the reaction. Each flavor group contains only a few particles, so the overall pattern of lifetimes within a flavor group is not entirely clear-cut.

Two striking features of these quark lifetimes are (a) the obvious α-quantization of the (u, d), s, b intrinsic lifetimes, and (b) the accurate experimental factor of 3 spacing between the b and c intrinsic quark lifetimes, as displayed in Fig. 2.5. A theoretical challenge which is posed here is to relate these quark lifetimes to their energy-quantized substates, which are listed in Table 3.1 of Chapter 3.

2.7   The α^4 Gap Between Flavor-Breaking and Flavor-Conserving Particle Decays

In the preceding sections we have discussed the separation of the elementary particle *unpaired-quark* ground states into discrete *flavor groups*, and the patterning of the lifetimes within these flavor groups. These long-lived unpaired-quark states decay by *flavor-breaking* electroweak interactions that greatly inhibit their decay rates. In addition to these particles, there are within each flavor group some much-shorter-lived *paired-quark–antiquark* and *radiative* ground states that decay by *flavor-conserving* interactions. These two types of ground-state particle lifetimes, flavor-breaking and flavor-conserving, are cleanly separated from one another by a characteristic factor of α^4 — eight orders of magnitude. The clean separation between these two types of lifetimes is an important phenomenological result. This separation is in the form of an "α^4 gap" — a "lifetime desert" — where no particle lifetimes are observed. There are no "rogue" particles in the systematics of this quark-dominated particle lifetime spectrum.

Figure 2.7 displays the lifetimes of all these low-mass ground states, which are the possible combinations of the u, d, s, c, b quark flavors, plotted together logarithmically on a common α-spaced lifetime grid. This global plot demonstrates the various ways in which the renormalized fine structure constant $\alpha \sim 1/137$ dominates the systematics of their decays. It is probably the most significant figure in Chapter 2.

The *hadronic* (strongly interacting) elementary particles occur in two different forms: (1) half-integer-spin *baryons* that contain three quarks; (2) integer-spin *mesons* that contain a quark and an antiquark. As can be seen in Fig. 2.7, and also in Fig. 2.4, the baryons and mesons within a flavor group have essentially the same lifetimes. Thus, the lifetime of a particle is *not* determined by the *geometry* of its quark combination, but simply by the *lifetime* of the *dominant quark* in the particle, as defined in Sec. 2.5.

As described above, the short-lived flavor-conserving and long-lived flavor-breaking particle lifetimes within a flavor group are separated from one another by four powers of α (eight orders of magnitude). Remarkably, this α^4 lifetime ratio is essentially the same for all of the flavor groupings. It is accurately a factor of α^4 for the pseudoscalar (ps) mesons, and approximately α^4 for the other flavor groups. Moreover, it shifts by a factor of α in going from the $x_i = 0$ (ps)-group to the $x_i = 1$ s-quark group, and by an accurate factor of α^2 in going from the $x_i = 0$ (ps)-group to the $x_i = 2$ b-quark group. Finally, it shifts by an accurate factor of 3 in going from the

b-quark group to the c-quark group. (This b-quark to c-quark lifetime factor of 3 is displayed in detail in Fig. 2.5, where the c-quark central lifetime $CL(c)$ has the value $x_i = 2.2229$, which is a displacement by a numerical factor of 2.994 from the $CL(b)x_i = 2$ α-grid line.)

As shown in Fig. 2.7, these shifted α^4 lifetime gaps combine together to create a vast *lifetime desert* where no non-conforming *rogue* particle lifetimes are observed.

The one common factor in the various flavor-breaking slow decays is that they are all electroweak decays that are mediated by the W and Z gauge bosons. This suggests that the characteristic α^4 gap is related to the W and Z mass values. This conclusion is consistent with the mass/energy systematics established in Chapter 3, which relates the production of the W and Z bosons to the fine structure constant α.

Figure 2.7 visually summarizes the main features of the quark ground-state lifetimes. The overarching feature that emerges from this display of lifetime patterns is the decisive role played by the $\alpha \sim 1/137$ renormalized fine structure constant. The factor of α appears in the values of the individual $(u, d), s, c, b$ intrinsic quark lifetimes, as shown by the positions of the separated lifetime flavor groups along the α-spaced abscissa. The factor of α^4 defines the constant width of the lifetime desert, which is maintained while the boundaries of this width are being shifted laterally by factors of α. And the lifetimes of the η and η' excited states of the π meson are accurately spaced by powers of α. Only by plotting all of these lifetimes together in a comprehensive display can we relate these various systematic features. And without the implementation of the scaling factor α, the behavior of these lifetimes would be very difficult to interpret. Figure 2.7 should be a central part of any phenomenological discussion of elementary particle lifetime systematics.

Lifetime alpha quantization and α^4 gaps

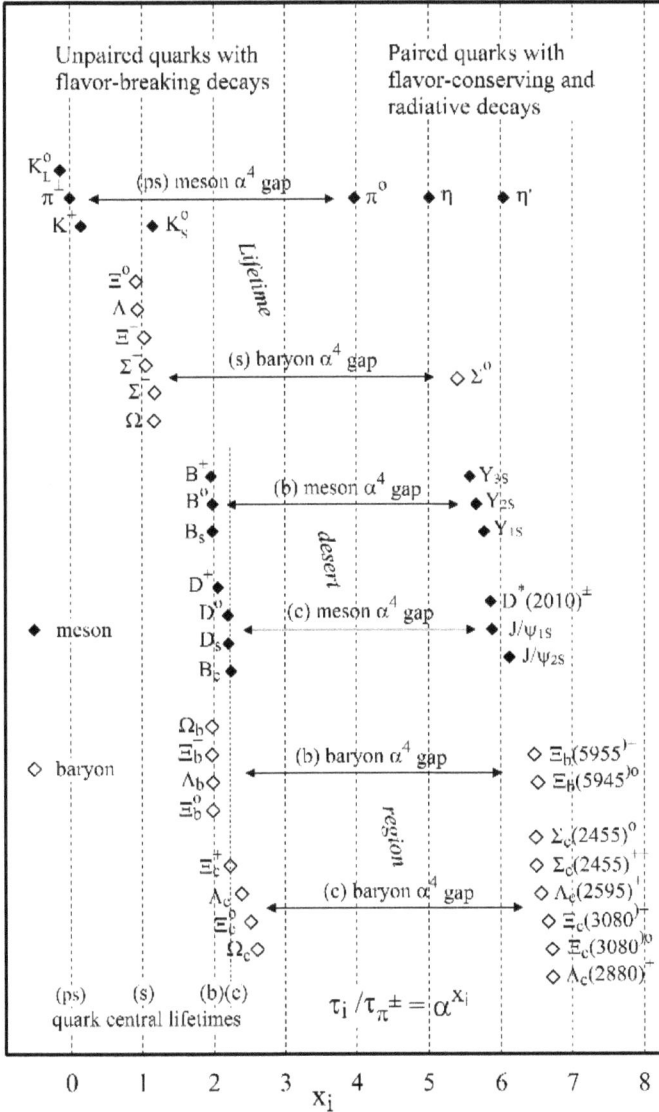

Figure 2.7 The ground-state particle lifetimes, shown divided into (ps), (s), (b), (c) lifetime flavor groups, and also into meson and baryon families. As can be seen, the flavor-breaking (weak-interaction) decays in each group are separated from the flavor-conserving decays by a factor of approximately α^4, thus creating a *lifetime desert region* in which no lifetimes occur.

2.8   The Average-Deviation-from-an-Integer (ADI) Minimization Test of Lifetime Scaling Factors

In the investigation of scaling factors S that apply to particle lifetimes, one of the tasks is to numerically investigate the accuracy to which the experimental lifetime data themselves define S. A standard method for accomplishing this is the $\chi^2(S)$ least-squares minimization technique described in Sec. 2.4. It is based on a *quadratic* fit to the experimental data points weighted by their experimental uncertainties, as defined in the $\chi^2(S)$ sum of Eq. (2.1). The results for a $\chi^2(S)$ fit to the $\pi^\pm$, π^0, η, η' data set were displayed in Fig. 2.3.

There is also a *linear* "absolute-deviation-from-an-integer" (ADI) technique that we can use to evaluate particle lifetime scaling $\tau_i(S)$ as a function of S. We can illustrate how ADI minimization works by applying it to the same $\pi^\pm$, π^0, η, η' nonstrange pseudoscalar meson data set that was selected for the χ^2 curve in Fig. 2.3. The $\pi^\pm$ meson is used as the reference lifetime, and the π^0, η and η' lifetimes τ_i are calculated as ratios to the $\pi^\pm$ lifetime by the equation $\tau_{\pi^\pm} S^{-x_i} = \tau_i(\text{exper})$, which determines the exponent $x_i(S)$ for the lifetime τ_i. If the equation for τ_i is exact, the exponent x_i will be an integer (perfect S scaling). In general, x_i is not an integer, and the deviation dx_i from an integer is a measure of the inaccuracy of the S scaling. The deviation that is relevant is the *absolute value* of the deviation, $\Delta x_i \equiv |dx_i|$. To evaluate the accuracy of the scaling factor S for a set of N data points, the *average absolute deviation from an integer* ADI(S) is calculated by the following equation:

$$\text{ADI}(S) = \frac{1}{N} \sum_{i=1}^{N} |\Delta x_i|(S), \quad \Delta x_i \equiv |x_i - I_n|, \quad I_n = \text{nearest integer}.$$

$$(2.5)$$

The scaling factor S that gives the best fit to the experimental data is the one that yields the minimum value for $ADI(S)$, which is obtained by plotting the function $ADI(S)$ over a range of S values. Since the maximum absolute deviation Δx_i of an individual lifetime exponent x_i from the nearest integer value I_n is 0.5, a set of random lifetimes that have *no* scaling factor will show a *random* distribution of x_i's that range between 0 and 0.5 and give the average absolute deviation value $ADI \cong 0.25$. But if the experimental lifetimes in the test set do exhibit a significant scaling regularity for a region of S values, then the $ADI(S)$ curve will contain a dip below 0.25 in this region.

Figure 2.8 displays the $ADI(S)$ curve for the $\pi^{\pm}$, π^0, η, η' data set over the scaling range $131 < S < 145$. Due to the small number of lifetimes, and the discontinuity in the second derivative as an integer value for x_i is passed, the slope of the $ADI(S)$ curve is not continuous. A minimum at

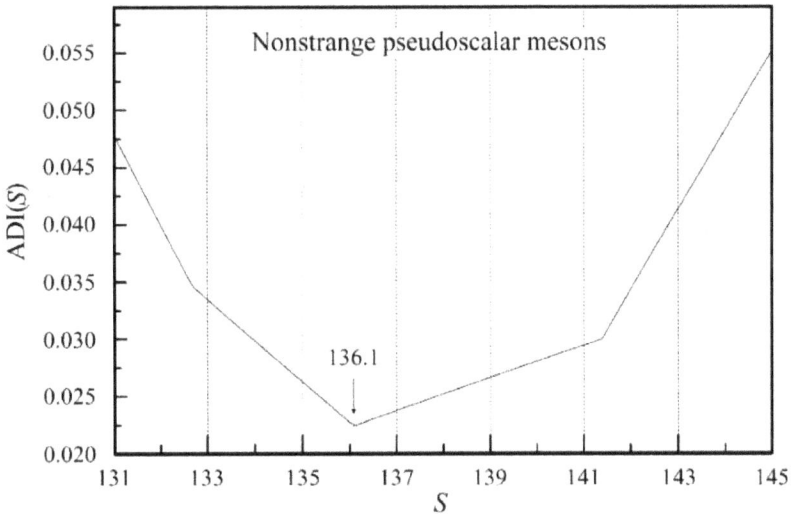

Figure 2.8 The $ADI(S)$ curve for the $\pi^{\pm}$, π^0, η, η' lifetime data set, which has a minimum value of 0.0225 at $S = 136.1$. This is close to the lifetime quantization value $1/\alpha \sim 137.04$ that is displayed for these data in Fig. 2.2a.

$S = 136.1$ is clearly in evidence. The value at the minimum is $ADI(S_{min}) = 0.0225$, which is much smaller than the value $ADI \cong 0.25$ that corresponds to a random fit. This verifies that S_{min} is a common scaling factor for these pseudoscalar meson lifetimes. Furthermore, the ADI value $S_{min} = 136.1$ is within 0.7% of the α-spaced value $S = \alpha^{-1} = 137.0$, and the χ^2 value $S_{min} = 139.1$ in Fig. 2.3 is within 1.5% of the value 137.0 in the other direction. Thus, these two *independent* ways of analyzing the lifetime scaling in factors of S yield S_{min} values that straddle the fine structure value $\alpha^{-1} = 137.0$, and they provide a quantitative evaluation of the accuracy of the pseudoscalar meson lifetime α-scaling that is visually apparent in Fig. 2.2a.

The $ADI(S)$ minimization test can also be used to demonstrate the difference between the lifetime patterns of the long-lived ground-state flavor groups and those of the corresponding short-lived excited states. The $ADI(S)$ curves for these two groups of particle states are plotted together in Fig. 2.9. As can be seen, the short-lived excited states of curve (a) have an oscillating distribution of ADI values centered on $ADI = 0.25$, which corresponds to a random distribution of lifetimes. The long-lived ground states of curve (b), on the other hand, exhibit a periodicity in powers of S that approximates the fine structure constant value $1/\alpha \sim 137$, as shown by the dip in curve (b) down to the value $ADI = 0.153$, which is centered at S_{min} is 133.4.

The ADI S_{min} value of 133.4 for 43 particle ground states is in essential agreement with the 4-particle ADI S_{min} value of 136.1 shown in Fig. 2.8 and the 4-particle χ^2 value of 139.1 shown in Fig. 2.3. Thus, it seems clear experimentally that the elementary particle ground-state lifetimes do exhibit a periodicity in the $1/\alpha \sim 137$ *renormalized* fine structure constant value. Also important is the fact that all of the hadronic ground-state lifetimes which been observed to date fit into the comprehensive α-spaced lifetime systematics that was first defined by the early lifetime measurements shown

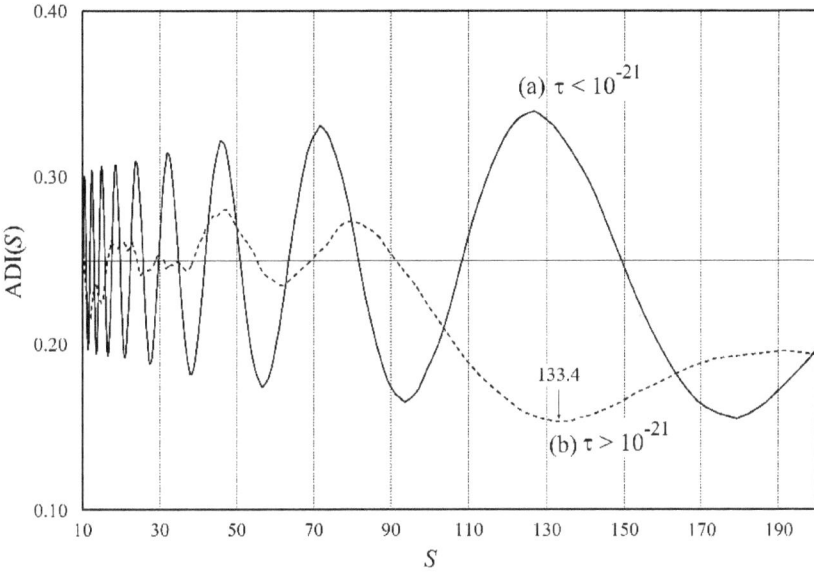

Figure 2.9 The *average absolute-deviation-from-an-integer* ADI(S) curves for the measured particle lifetimes (App. B). Curve (b) is for the long-lived ground-state particles of Zones 1–3 in Fig. 2.1, which are the unpaired-quark decays at the left in Fig. 2.7. Curve (a) is for the short-lived excited-state particles of Zone 4 in Fig. 2.1. The ADI curve (a) for the excited-state particles oscillates around the ADI = 0.25 value, which denotes no periodic scaling of the particle lifetimes. The ADI curve (b) for the ground-state particles, on the other hand, shows a dip as the scaling factor S approaches the value $1/\alpha \sim 137$. The minimum at $S_{min} = 133.4$ has the value ADI = 0.153, which is statistically much smaller than the random value 0.25.

in Fig. 2.2, and later developed into the full-fledged ground-state mapping displayed in Fig. 2.7.

Historically, lifetime ADI data analyses were published in 1974 [5], just before the J/Ψ particle discovery, and in 1976 [6], with the J/Ψ, Ψ(2S), χ(1P), and η(2S) lifetimes added in. They strengthen the present ADI-based conclusions as to the relevance of the *renormalized* fine structure constant $1/\alpha \sim 137$ in creating formalisms to calculate both hadronic and leptonic ground-state particle decay processes.

2.9　The n-to-$\mu(\alpha^{-4})$ and τ-to-$\mu(\alpha^3/3)$ Lifetime α-Quantized Ratios

The α-quantized lifetime studies in the preceding sections have involved the *hadronic* quark and particle ground states. Theoretically, it is not clear that the *leptonic* muon and tauon lifetimes should bear any relationship to the hadronic lifetime α-quantization, and it is also not clear that the *hadronic* neutron and *leptonic* muon lifetimes should be related to one another. However, as displayed in Fig. 2.10, the neutron, muon and tauon lifetimes seem to be phenomenologically related in an interesting manner, as we now discuss.

The muon lifetime is shorter than the neutron lifetime by a factor of approximately α^4, as shown in Fig. 2.10. This α^4 lifetime spacing seems

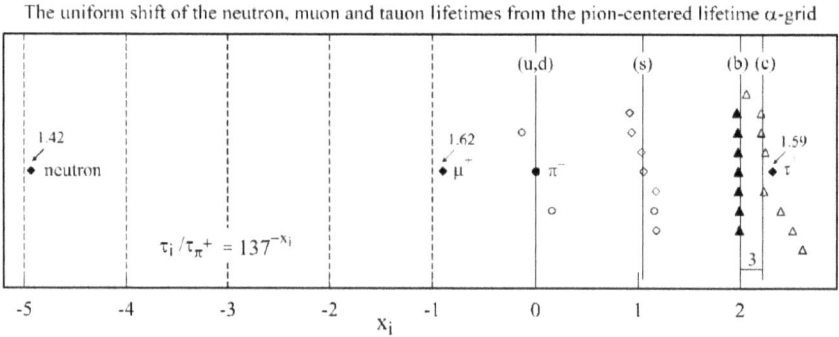

Figure 2.10　The (u,d), (s), (c) and (b) ground-state lifetime flavor groups of Fig. 2.5 are displayed here together with the neutron, muon and tauon lifetimes, using the same α-spaced lifetime grid anchored on the $\pi^{\pm}$ lifetime. The solid vertical lines in the four flavor groups denote their central lifetimes CL. The neutron (factor of 1.42), $\mu^{\pm}$ (factor of 1.62) and $\tau^{\pm}$ (factor of 1.59) lifetimes do not fall directly on the α-grid lines, but are all similarly shifted slightly towards shorter lifetime values. The neutron and muon shifts are with respect to the α-grid itself. But the tauon shift is with respect to the (c) flavor-group CL, which itself is shifted by a factor of 3 toward shorter lifetimes. These similar shifts suggest that these three lifetimes may be numerically linked together via a slightly shifted α-grid that is anchored on the $\mu^{\pm}$ lifetime.

analogous to the six α^4 lifetime gaps displayed in Fig. 2.7, which separate the long-lived *flavor-breaking* lifetime flavor groups from their shorter-lived *flavor-conserving* counterparts. The flavor-breaking process, which involves the transformation of an unpaired quark, is an electroweak transition that is mediated by the W and Z gauge bosons, whose large masses dictate the magnitude of the α^4 gaps shown in Fig. 2.7.

A similar flavor-breaking and flavor-conserving situation applies to the neutron and muon lifetimes, which have the following ratio:

$$\tau_{\text{neutron}} = \tau_{\text{muon}} \times \alpha^{-4} \ (14\% \ \text{accuracy}). \tag{2.6}$$

The neutron and muon decay modes are

$$\text{neutron}: \ d \to u + e + \bar{v}_e; \quad \text{muon}: \ \mu \to v_\mu + e + \bar{v}_e. \tag{2.7}$$

The neutron decay involves the *flavor-breaking* transition from a d to u quark, together with the *lepton-conserving* creation of an $(e, \bar{v}_e)$ pair. The muon decay involves the transition from a μ lepton to a v_μ muon neutrino, together with the creation of an $(e, \bar{v}_e)$ pair, both of which are *lepton-conserving* processes, and do not involve the flavor quantum number. The required d to u flavor-breaking electroweak transformation in the neutron decay channel increases its mean lifetime by a factor of $\alpha^{-4} = 3.53 \times 10^8$ relative to that of the muon. Experimentally, the neutron to muon lifetime ratio is 4.01×10^8, which differs from the eight orders of magnitude α^{-4} electroweak decay factor by only 14%. The relative closeness of this agreement indicates that hadrons and leptons share the same basic α-quantized lifetime formalism, as is suggested by the lifetime systematics displayed in Fig. 2.10.

The lifetime ratio of the tauon and muon leptons introduces two novel features into the α-quantized lifetime systematics. Their lifetime ratio is

$$\tau_{\text{tauon}} = \tau_{\text{muon}} \times \alpha^3/3 \text{ (2.0\% accuracy)}. \qquad (2.8)$$

The scaling factor of $\alpha^3 = 2.53 \times 10^{-6}$ between these two leptons is unique, and is not found elsewhere in the particle lifetime systematics. The additional scaling factor of $1/3$ is in line with the difference between the central lifetimes CL(b) and CL(c) in Fig. 2.10. The c-quark flavor group is shifted toward shorter lifetimes by a factor of 3 relative to the b-quark flavor group. The surprising result here is that the *leptonic* tauon has the same factor of $1/3$ lifetime shift that is observed in the *hadronic* c-quark particle lifetimes. In the energy relationships between these particles discussed in Chapter 1, it was shown that the s, c and b quarks form an energy-tripling sequence in which the c-quark energy is 3 times the s-quark energy, and the b-quark energy is 3 times the c-quark energy. Since the quark energy and lifetime are conjugate quantities, it is perhaps not surprising that the *lifetimes* of the c-quark states differ from the lifetimes of the b-quark states by a factor of 3, which reflects the factor-of-3 difference in their *energy* substructures (although it is not clear why the c-quark states should have the shorter lifetimes). But the tau lepton has no discernable mass/energy substructure, so the fact that its lifetime also contains this factor of 3 shift is puzzling. The 2% accuracy of the overall τ-to-μ ($\alpha^3/3$) scaling factor, which spans 6 orders of magnitude, suggests that this result is not accidental. Equation (2.8) is one of the most significant empirical results to emerge from the α-quantized lifetime phenomenology.

The elementary particle α-quantized energy formalism discussed in Chapter 1 and the α-spaced lifetime regularities displayed in Chapter 2 both have hadronic and leptonic particles combined together on an equal footing. These results should be helpful in determining the essential relationships among the various particle families, and they underscore the *conjugate* nature of particle energies and lifetimes.

2.10   The Evolution of the Lifetime α-Grid from 1970 to 2018

One of the most important aspects of elementary particle *lifetime α-quantization* is the historical manner in which it appeared in the experimental data. The early particle detectors had a mass/energy accuracy of about $1\,\mathrm{MeV}$, which corresponds to a particle lifetime of about 1.5×10^{-21} sec. Thus, only relatively-long-lived particle lifetimes could be determined. Also, the accelerator energies were not high enough to generate the full spectrum of energy states that we know about today. In spite of these limitations, the α-quantization of particle lifetimes was already clearly apparent by 1970, as is displayed in Fig. 2.11.

The bottom section of Fig. 2.11 shows the elementary particle lifetime data base as it existed in 1970: five mesons $(\pi^\pm, \pi^0, \mathrm{K}^\pm, \mathrm{K}^0_\mathrm{L}, \mathrm{K}^0_\mathrm{S})$, seven baryons $(n, \Lambda^0, \Sigma^-, \Sigma^+, \Xi^0, \Xi^-, \Omega^-)$, and the $\mu^\pm$ lepton — thirteen lifetimes in all. If the $\pi^\pm$ lifetime is selected as the reference unit lifetime, and if the other particle lifetimes are expressed as logarithms x_i to the base α, where α is the renormalized fine structure constant $\alpha = e^2/\hbar c \cong 1/137$ then all 13 of these particle lifetimes are closely grouped along the α-spaced logarithmic grid lines, as displayed at the bottom of Fig. 2.11. As an historical note, the calculations of the lifetime logarithm x_i values were initially carried out, in those pre-computer days [7], on a Marchant mechanical calculator.

Theoretically, the challenge to the α-spaced lifetime grid that was established by the 1970 lifetime data was to see how it accommodated the later-discovered ground-state quarks and leptons. The 1970 meson and baryon states are composed of u, d and s quark combinations. They are mainly

The α-quantized particle lifetime spectrum in 1970 and in 2018

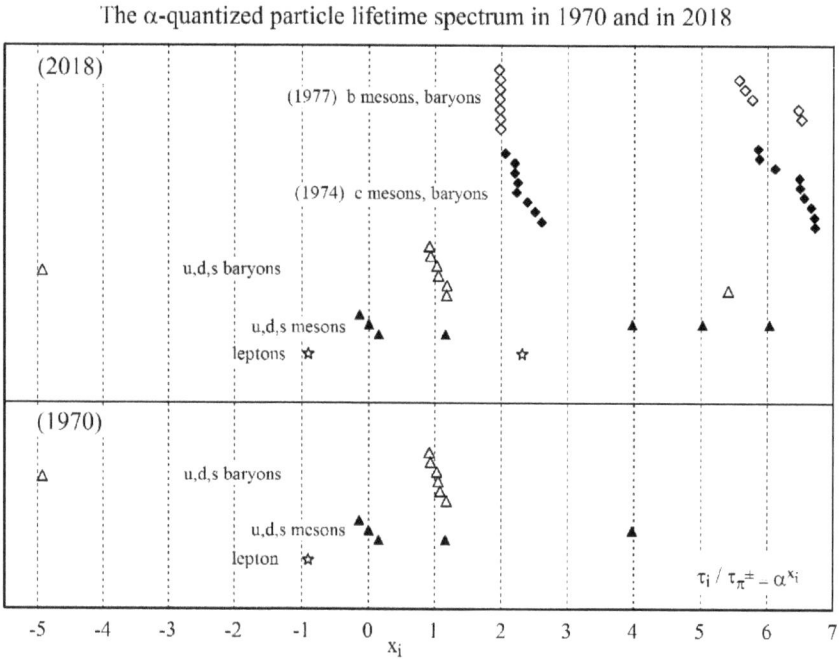

Figure 2.11 The experimental α-quantized elementary particle lifetime data base as it was in 1970 (bottom data plot), and in 2018 (top data plot). These are the lifetimes of the long-lived hadron quark-flavor ground states and the lepton states. The α-grid for Fig. 2.11, which is anchored on the $\pi^{\pm}$ lifetime, was determined by the 13 lifetimes known in 1970 (which are shown here with their original values), and it accurately applies to the 33 particles with lifetimes longer than $\tau > 10^{-21}$ sec that have subsequently appeared. These include the c-quark (1974) and b-quark (1977) states, which form the separated c and b flavor groups displayed here in Fig. 2.11, and in more detail in Fig. 2.7. It is noteworthy that no rogue lifetimes have appeared.

grouped along the $x_i = 0$ and $x_i = 1$ logarithmic abscissas. The c quark mesons and baryons that emerged starting in 1974 are grouped somewhat beyond the $x_i = 2$ abscissa, which made their α-quantization difficult to evaluate. However, the b quark mesons and baryons, which emerged in 1977, are spectacularly grouped precisely along the $x_i = 2$ abscissa, as shown in the 2018 data plot of Fig. 2.11. Hence, their α-quantization seems experimentally unassailable. Furthermore, the ground-state c-quark mesons and baryons have a central lifetime (CL) grouping that is an accurate factor of 3 shorter than the b-quark CL grouping (Fig. 2.5). In the absence of a comprehensive theory for these ground-state flavor-breaking lifetime decays, these results are purely phenomenological, but the fact that the fine structure constant α plays a key role seems undeniable. The only lepton that has emerged since 1970 is the tauon, and it fits into the c-group of hadrons with respect to both its lifetime and its mass/energy value.

In the 48 years from 1970 (when the lifetime α-grid was first published [7]) to 2018, 33 new long-lived particle states emerged, as illustrated in the bottom section of Fig. 2.11. These include the $\tau^{\pm}$ lepton, 3 (u, d, s) hadrons, 17 c-group hadrons, and 12 b-group hadrons. The 8 c-group and 7 b-group long-lived hadrons are the ground states for the various quark combinations. They represent an essentially complete mapping of the possible quark-flavor combinations, which means that the particle physics experimentalists have completed the task of measuring the hadron ground-state lifetimes. This implies that the basic lifetime regularities displayed in the top section of Fig. 2.11 are not going to change significantly in the future. In particular, the fact that no *rogue* ground-state lifetimes have appeared — isolated lifetimes that lie well off the established α-grid — is an argument for the relevance and comprehensiveness of the elementary particle lifetime α-quantization phenomenon.

The historical expansion of the elementary particle lifetime data base is illustrated in Table 2.3. This expansion occurred as the result of increases in the available accelerator energies and improvements in the particle detector efficiencies. This mature data set is essentially the final one that particle theorists will have to work with.

Table 2.3 The expansion of the experimental lifetime data base in the 48 years from 1970 to 2018. The observed α quantization of particle lifetimes occurs in the long-lived flavor-breaking decays. This result is not a feature of the current Standard Model formalism.

Year	$\tau_{\text{mean}} > 10^{-21}$ sec (Ground states)	$\tau_{\text{mean}} < 10^{-21}$ sec (Excited states)	τ_{mean} total
1970	13		
1974	13		
1976	17		
1990	28	101	129
2005	36	121	157
2018	47	166	213

2.11   Summary and Outlook

The elementary particle lifetime analysis carried out in Chapter 2 illustrates the α-quantization of the long-lived particle ground states, and it also demonstrates that the α-quantized lifetimes of the hadronic states are carried by the quarks themselves. These are stand-alone results that needs no outside confirmation for their validity. However, reinforcement for these results is obtained from the fact that the particle lifetimes are the quantum mechanical conjugates of the particle energies, which are shown in Chapter 3 to be not only α-quantized, but α-generated in α-*boost* excitations from well-defined ground states. This α-boost excitation process involves the creation of quantized *energy packets*, which combine together to form the quark and particle energies. Thus, the particle quarks have an α-quantized energy substructure (at least conceptually) that is reflected in the α-quantized quark lifetimes, although the details of this relationship remain to be worked out.

In Chapter 4 we delve inside the fine structure constant $\alpha = e^2/\hbar c$ itself and obtain energy relationships that shed further light on the particle generation process, including a dramatic phase transition that occurs during the conversion of particle coulomb or kinetic *energy* into particle non-electromagnetic *mechanical* mass.

References

[1] M. Tanabashi *et al.* (Particle Data Group) *Phys. Rev. D* **98**, 030001 (2018) (URL: http://pdg.lbl.gov).

[2] R. A. Arndt and M. H. Mac Gregor, "Nucleon-Nucleon Phase Shift Analyses by Chi-Squared Minimization," in *Methods in Computational Physics, Vol. 6. Nuclear Physics.*, B. Alder, S. Fernbach and M. Rotenberg, eds. (Academic Press, New York, 1966), pp. 253–296.

[3] R. P. Feynman, *QED: The Strange Theory of Light and Matter* (Princeton University Press, 1985), p. 129.

[4] J. D. Jackson, *The Physics of Elementary Particles* (Princeton University Press, 1958), pp. 57–58.

[5] M. H. Mac Gregor, "Experimental Systematics of Particle Lifetimes and Widths," *Il Nuovo Cim. A* **20**, 471–507 (1974), Figs. 3 and 4.

[6] M. H. Mac Gregor, "Lifetimes of SU(3) Groups and J Particles as a Scaling in Powers of α", *Phys. Rev. D* **13**, 574–590 (1976), Fig. 4.

[7] M. H. Mac Gregor, *Lett. Nuovo Cim.* **4**, 1309–1315 (1970), Table V.

Chapter 3   The Experimental α-Quantization of Elementary Particle Energies and Masses, where $E = mc^2$

3.1   Introduction

The field of elementary particle physics is devoted to the study of the basic constituents of matter. For the past century, it has been one of the most intriguing areas of physics. The development of large particle accelerators and particle detectors has enabled physicists to go to high particle energies, and consequently to probe into short distances, where matter can yield the secrets of its smallest units — the particles that make up atoms and molecules. An imposing data base of particle properties has been compiled [1]. There are indications that this data base may be nearing its final form, at least with respect to the basic features of the elementary particle spectrum. Thus, the experimental particle physicists may have completed their mission of locating and measuring the various elementary particle states. However, one of the primary goals in this mission — to understand the observed mass *patterns* in the particle data base — has not yet been accomplished.

Perhaps an even more important goal for particle physics is to understand the *nature of particle mass* itself. An elementary particle, in its basic essence, is a localized quantity of energy. As has been recognized for a long time, the energy of the electron, or the proton, is not purely electromagnetic. There needs to be a non-electromagnetic *mechanical mass* that localizes and stabilizes the particle. But the experiments have not yet revealed the properties of this mechanical mass. Also, turning to cosmology, the spiral galaxies seem to contain *dark matter* that is necessary to hold them together gravitationally. And the universe itself seems to be expanding under the impetus of a *dark energy* — Einstein's cosmological constant. According to the latest cosmological models, the visible matter in the universe — the stars and galaxies — represents 4% of its total energy, the dark matter represents 20%, and the other 76% is dark energy. The relationships among these three energy components in the universe have yet to be established.

If the particle data base is indeed essentially complete, then the task becomes one of poring over the data to see if something has been overlooked. It may be that if we can find an enigma — a puzzling data relationship — hidden somewhere in the data, it will lead to other discoveries. As we will see, this turns out to be the case. And, rather surprisingly, this enigma occurs at the highest energies in the data base.

3.2   The Elementary Particle Data Base

The particle data base for the present studies is the set of 213 particles that have well-determined masses and lifetimes, as compiled in the current Review of Particle Properties (RPP) [1]. Particle masses m and their corresponding energies E are related by the Einstein equation $E = mc^2$, where c is the velocity of light. In the present work we will deal mainly with energies rather than masses. Particle energies are conjugate to particle lifetimes. The particle energies are listed in Appendix B.

In the 213-particle data set, only two of the particles are stable — the electron and proton. The unstable particles include two leptons, 34 long-lived hadron *ground states* of the various u, d, s, c, b quark flavor combinations with lifetimes greater than 10^{-21} sec (1 yoctosec), and 177 short-lived hadron *excited states*. The mass spectrum of these 213 states extends up to 11 GeV. Above this energy there is a large 70 GeV particle void. Then four more particle states appear — the W and Z gauge bosons, the Higgs boson, and the top quark t, with energies of 80, 91, 125, and 173 GeV, respectively. No particle states above 173 GeV have been detected. Since the top quark t must be produced in matching $(t, \bar{t})$ pairs in order to preserve the t-quark flavor number, a $t, \bar{t}$ resonance at 346 GeV might be expected to appear, but nothing is observed there. This is because the t lifetime is so short (a few yoctoseconds) that there is no time for the t quark pair to bind together. In the LHC p–p collisions, the impact excitation energy that is available for particle formation extends up to roughly 4 TeV, but no particles other than those mentioned above have been detected.

The elementary particle energy region up to 11 GeV has been thoroughly explored, and all of the expected hadronic combinations of the u, d, s, c, b quarks have been detected. In the searches for the elusive Higgs boson, the energies above 11 GeV have also been extensively investigated. In addition to the particle energy void from 11 to 80 GeV, the energy region above the 173 GeV energy of the top quark t has thus far turned out to be devoid of particles. Particle and quark lifetimes get very short at these high energies — shorter than the time required for quark hadronization. Thus, a natural excitation energy upper limit may exist for elementary particle formation above 173 GeV. The particle data base that we now have may be complete, at least with respect to its most basic particle properties.

3.3 The Generation of a Mass Spectrum from a Basis Set of Stable Ground-State Particles

In nuclear physics, the classic example for the generation of the mass spectrum of related particles is the set of atomic nuclei, whose basis set is the lowest-mass stable proton and bound neutron. The *atomic number* A is the number of protons in the nucleus, and the atomic mass spectrum extends from A = 1 to 92, with a further extension to observed unstable atoms above A = 92. These atoms are generated as multiples of the proton and neutron ground states.

When this mass generation mechanism is applied to the elementary particle spectrum, it expands to include quarks, leptons and baryons. The particle masses are related to the particle energies by the Einstein equation $E = mc^2$. Since the particle energies are conjugate to the lifetimes discussed in Chapter 2, we will study the particle generation process in the *energy representation*, which enables us to include all types of particle in a single excitation pattern.

The particle generation process is illustrated in Fig. 3.1, which displays the α-quantized energy generation mechanism for the non-strange pseudoscalar π, η and η' mesons. The ground-state is the 1.022 MeV electron-positron particle-antiparticle pair. The first excited state is the 139 MeV spin 0 $\pi^\pm$ meson, which is its own antiparticle. The factor of 137 α-*leap* from the $e^- e^+$ ground state generates a 137 MeV *energy packet*, which is then multiplied by successive energy leaps so as to accurately reproduce the energies of the η and η' mesons.

The stability of these mesons is expressed in Fig. 3.2 as the α-quantized logarithmic decay rates for the π, η and η' mesons. The slow decay rate of the 139 MeV $\pi^\pm$ mesons is due to the fact that they have unpaired-quark (electroweak) decays that do not conserve the quark flavors. The π^0, η and η' mesons decays involve paired-quark decays that do conserve the quark flavors. The crucial fact here is that these paired-quark decays are accurately α-quantized, as shown in Fig. 3.2.

Figure 3.3 displays the pseudoscalar meson energies, expanded to include the K mesons. These states are generated by energy packets of

The α-leap energy-packet quantization of
the non-strange pseudoscalar mesons

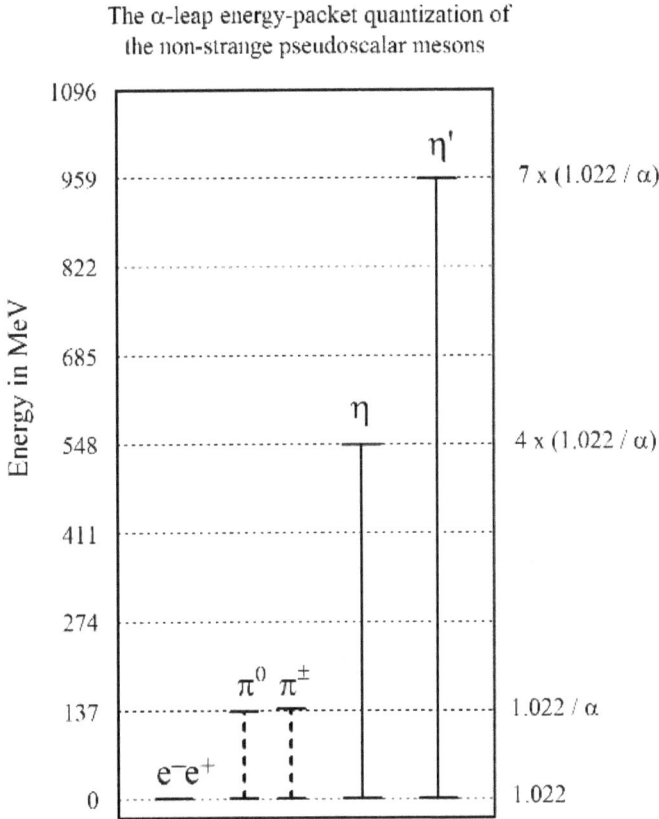

Figure 3.1   The generation of the non-strange scalar mesons by repeated α-leaps from the electron–positron pair that generate 137 MeV energy packets, which then combine together as shown in multiples of 1, 4 and 7 to accurately reproduce the observed particle energies.

70.025 MeV. The eta meson has 8 energy packets, or 560.2 MeV. The total energy available for eta meson production is 560.2 MeV plus the electron energy of 1.022 MeV, or 561.22 MeV. The experimental energy is 547.862 MeV. Thus, the binding energy we ascribe to the eta meson is B.E. $= 547.862/561.22 = 0.975$, or 2.5%.

The η' meson requires 14 energy packets, or 980.4 MeV. The total energy available for η' meson production is 980.4 MeV plus the electron

The α-quantized decay rates d of
the non-strange pseudoscalar mesons

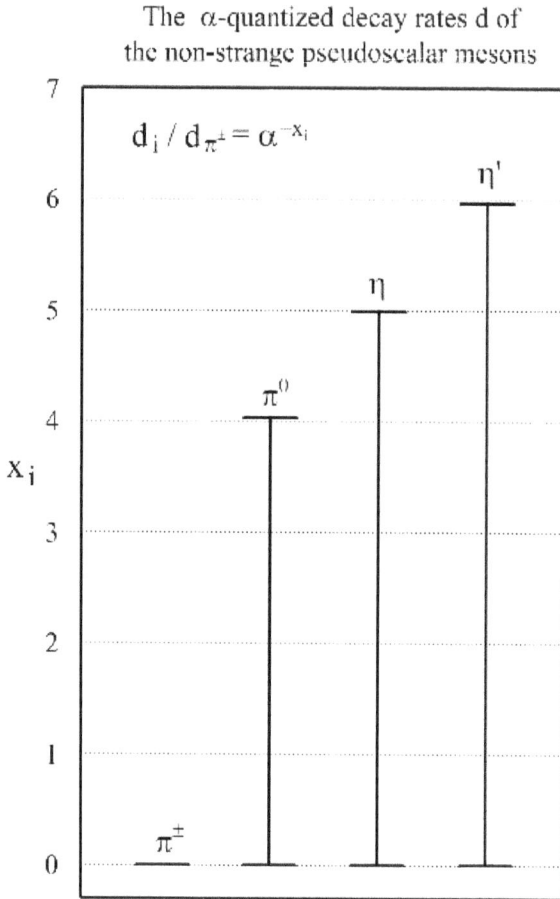

Figure 3.2   The α-spaced logarithmic-energies of the π, η and η' pseudoscalar mesons indicate stability from the decay rates of the non-strange scalar mesons by repeated α-leaps. The long lifetime of the $\pi^{\pm}$ mesons follows because the decay of the unpaired quarks required an electroweak transformation that does not conserve the quark flavors. As can be seen in Fig. 3.2, the paired-quark decays are accurately α-quantized.

energy of 1.022 MeV, or 981.42 MeV. The experimental energy is 957.78 MeV. Thus, the binding energy we ascribe to the eta prime meson is B.E. $= 957.78/981.42 = 0.9756$, or 2.4%. The eta prime η packets themselves are spin 0 packets. The K meson available energy is $7 \times 70.025 = 490.175$

The pseudoscalar meson energies
as quantized in units of 70 MeV
α-generated energy packets

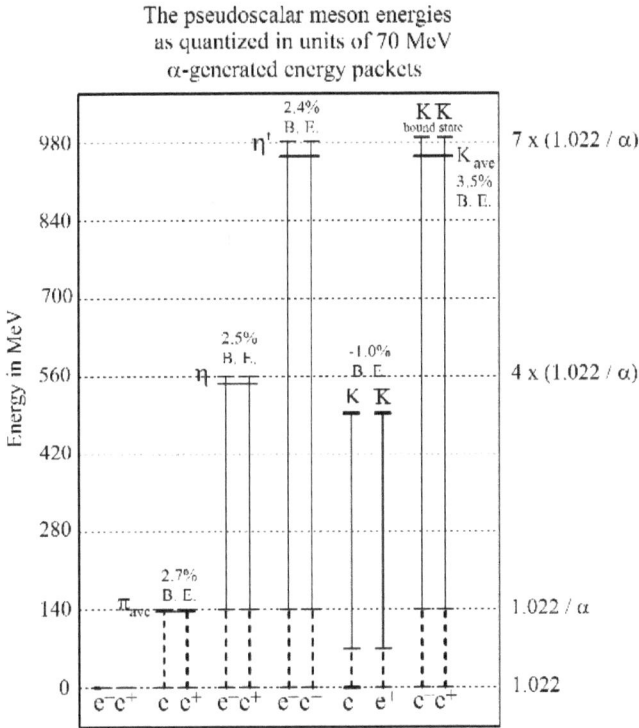

Figure 3.3 The pseudoscalar mesons, expanded to include the K mesons, and with the calculated binding energies (B.E.) displayed.

MeV plus the electron energy of 0.511 MeV, or 490.69 MeV. The kaon average energy is 495.66 MeV. Thus, a binding energy of $495.66/490.69 = 1.010$, or -1.0%, is ascribed to the K meson.

The eta prime meson, considered as a kaon–antikaon bound state, has 14 energy packets, or 980.35 MeV, plus the electron energy of 1.022 MeV, or 981.37 MeV. The kaon bound-state experimental energy is 957.78 MeV. Thus, the binding energy we ascribe to the kaon bound state is $957.78/981.37 = 0.9765$, or 3.5%.

The spin 0 bosons are the π, η and η' pseudoscalar mesons. They are generated by 70 MeV α-leaps from the electron ground state. The π^0 decay

The α-quantized energy packet production channels
from the electron to the spin 0 pions and spin 1/2 muons
and their decay channels back down to the ground states.

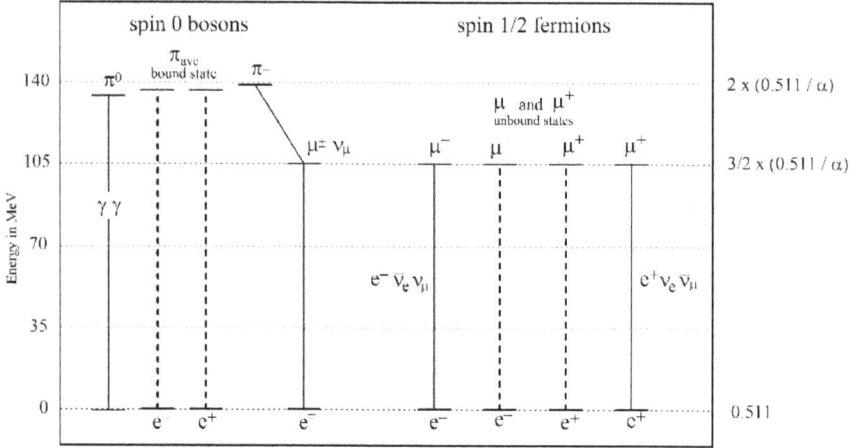

Figure 3.4 The α-quantized production and decay channels for the spin 0 bosons and spin 1/2 fermions.

is back down to the ground state, and it produces a characteristic α^{-4} gap in energy. The $\pi^{\pm}$ decay is to the spin 1/2 μ lepton, and lepton number must be conserved. So, the decay is to a lepton neutrino μ_ν plus an electron $e^{\pm}$ and an electron neutrino $\pm\nu_e$. What is significant here is that the η and η' meson lifetimes are accurately quantized in powers of $\alpha \sim 1/137$, as displayed in Fig. 2.2c above.

3.4   The Unexpected LHC Mass/Energy Equality
W + Z = Top Quark *t*

The CERN Large Hadron Collider (LHC) is one of the most impressive scientific projects ever undertaken. It took 11 years to build, and has had a total construction and operation cost to date of roughly 10 billion dollars. It serves a large pool of physicists from institutions all over the world. Early high energy experiments at CERN and Fermilab had already confirmed the predicted properties of the W and Z gauge bosons and top quark t. But particle physicists had several other LHC goals in mind: (1) to locate the elusive Higgs particle; (2) to find predicted supersymmetric partners of the existing particle spectrum; (3) to detect the axions that may play a role in dark matter; (4) to reveal unexpected "surprises" that could challenge existing theoretical concepts and lead to new avenues for exploration.

The discovery of a Higgs-like enhancement at 125 GeV fulfilled the first LHC goal. The Higgs peak is a manifestation of a scalar Higgs field, which supplies energy for particle creation. But it should be kept in mind that this Higgs formalism does not in any obvious way provide the formalism for determining what the particle mass values should be.

The second and third LHC goals, to discover the supersymmetric partners of the 39 particle ground states, and to find axions, have *not* been achieved. But the fourth goal, to unearth particle results that were not predicted, and that may lead to revisions in the way we regard these particles, has been realized. This unexpected result is an experimental relationship among the three highest-mass particle states, and it has extensive consequences that extend down to the lowest-mass particle states.

The four high-energy particle states are the gauge bosons $W^\pm$ and Z^0, the Higgs scalar boson H^0, and the top quark t. The Higgs boson has been identified by the ratios of its partial decay modes, which do not yield the precise mass values that have been obtained for the W, Z and t. The currently-listed Particle Data Group mass values in GeV/c^2 for these four states are [1]

$$W = 80.379; \ Z = 91.1876; \ H = 125.18; \ t = 173.0. \tag{3.1}$$

As these mass values demonstrate, there is a surprisingly accurate numerical mass equality between the (W, Z) gauge boson pair and the supposedly unrelated top quark t:

$$m_\text{W} + m_\text{Z} \equiv 2m_{\overline{\text{WZ}}} = m_t \ (0.99\% \text{ accuracy}), \tag{3.2}$$

where $m_{\overline{\text{WZ}}}$ denotes the *average mass* of the (W, Z) pair. The accuracy of this equality indicates that it may not be accidental. However, the $\text{W}^\pm$ and Z^0 gauge bosons (with integer spins, integer charges, and zero color) and the top quark t (with half-integer spin, fractional charge, and non-zero color) are totally different types of particles, and the Eq. (3.2) equality clearly does *not* apply to their electric charges, their spin angular momenta, or their color charges, which are *not* conserved. The one property these particle states have in common is their energy content E, where $E = mc^2$. Thus for Eq. (3.2) to be physically meaningful, it should be recast in the form of an energy equation, namely

$$m_\text{W}c^2 + m_\text{Z}c^2 \equiv 2m_{\overline{\text{WZ}}}c^2 = m_t c^2 \ (0.99\% \text{ accuracy}), \tag{3.3}$$

or in a simpler notation,

$$E_\text{W} + E_\text{Z} \equiv 2E_{\overline{\text{WZ}}} = E_t \ (0.99\% \text{ accuracy}). \tag{3.4}$$

Equation (3.4) is an empirical statement about intrinsic particle energies E. Its importance depends on the information that can be deduced from it with respect to particle formation [2]. It should be pointed out that part of the interest in Eq. (3.4) is the fact that it involves the experimental energy E_t of an individual unbound quark t. The energy E_t corresponds to its inertial mass m_t, which in Standard Model terminology is its *constituent-quark* mass. Thus, the LHC high-energy experiments offer a unique opportunity to study reactions that include an unbound freely-moving quark, since the experiments at low energies all involve quarks which are bound inside mesons or baryons, so that their intrinsic masses and energies cannot be directly determined. The results that emerge from these experiments are very informative [3].

3.5 The α-Boost Gauge-Boson Energy Packets E_{gb}

The *average gauge boson energy* $E_{\overline{WZ}}$ that appears in Eqs. (3.3) and (3.4) can be regarded as the empirical gauge boson *normalization* energy, whose experimental value is

$$E_{\overline{WZ}} \equiv \frac{(E_W + E_Z)}{2} = 85.786 \, \text{GeV}. \tag{3.5}$$

Superimposed on this normalization energy is the E_Z/E_W gauge boson *energy ratio*, which is independently specified by the electroweak equation $E_Z = E_W/\cos(\theta_W)$, where θ_W is the *weak mixing angle* [4]. In order to understand the significance of the normalization energy $E_{\overline{WZ}}$, there is an additional LHC-related piece of experimental information that should be introduced here: namely, the fact that we know the *ground state* energy in the colliding p–p beams, and can compare it to the *normalization* energy of Eq. (3.5). At these TeV energies, the proton is flattened relativistically in the direction of its velocity, and its three *up* and *down* valence energy quarks are essentially independent of one another. They are components of the well-studied parton energy distribution inside the proton. The p–p collisions at these high energies do not primarily involve the whole proton, but rather the individual components of the parton sea. The u and d valence quarks in the low-energy limit correspond to the u and d *constituent quarks* of the Standard Model [4]. A key experimental fact which applies here is that only one in every 10^{10} p–p collisions produces top quarks. The empirical interpretation offered for these rare events is that they represent head-on (valence quark)–(valence quark) u and d collisions, in which the quark pair absorbs enough of the total collision energy to create gauge bosons and top quarks. The d quark energy is slightly greater than the u quark energy, but this energy difference can be shown to average out when applied to the actual p–p quark collisions [5], so that the colliding u an d quarks function as equal-mass proton *energy quarks* $q_p \equiv (u, d)$, which each have the *effective* energy value

$$E_{q_p} = E_u = E_d = \frac{E_p}{3} = 312.757\,\text{MeV}. \tag{3.6}$$

A key point in defining the ground state of the p–p collisions is that the collision center-of-mass frame is also the rest-mass frame for the colliding valence quark pair q_p–q_p at the instant where they come into contact and their kinetic energy is about to be transformed into excitation energy. Thus, the *ground-state energy* of the q_p–q_p quark pair at this instant of time is the parton sea in its low-energy limit, where the energy of the proton resides in its three valence (constituent) energy quarks. Hence the collision ground-state energy is

$$E_{q_p} + E_{q_p} = \frac{(E_p + E_p)}{3} = 625.515\,\text{MeV}. \tag{3.7}$$

The LHC particle and quark excitations are produced from this ground-state energy base.

The calculation of the ratio of the *average gauge boson energy* of Eq. (3.5) to the colliding-beam q_p–q_p *ground-state energy* of Eq. (3.7) yields one of the most unanticipated and potentially important results to emerge thus far from the LHC high-energy particle experiments. This particle excitation energy-ratio, which is expressible in terms of the experimental p, W, and Z energies, is given by the equation

$$\frac{E_{\overline{WZ}}}{(E_{q_p} + E_{q_p})} = \frac{\left[\frac{(E_W + E_Z)}{2}\right]}{\left[\frac{(E_p + E_p)}{3}\right]} = 137.145. \tag{3.8}$$

The significance of this energy-ratio is that it almost exactly matches the value $1/\alpha$, where α is the renormalized *fine structure constant*, which is defined by the equation

$$\alpha = \frac{e^2}{\hbar c} \cong \frac{1}{137.036}. \tag{3.9}$$

The agreement between the factors of $\sim$137 displayed in Eqs. (3.8) and (3.9), which are both defined by experimental data, is to a numerical accuracy of 0.08%. This indicates that the *renormalized* (low-energy) constant α [6] plays a hitherto unsuspected role in the excitation mechanism for producing high-energy elementary particles. It is convenient to denote the factor of 137 increase in energy that occurs in going from the q_p-q_p quark-quark ground state energy of Eq. (3.7) to the $E_{\overline{WZ}}$ average gauge boson energy of Eq. (3.5) as an "α-boost" in energy. We can reinforce this result by applying it also to the production of low-energy pseudoscalar mesons, where the energy α-boost in going from the (e^-, e^+) electron–positron ground state to the $(\pi^\pm, \pi^0)$ pion pair is also a factor of 137, with no intervening particle states. The experimental data for these two examples are plotted together in Fig. 3.1.

The $(W^\pm, Z^0)$ *gauge bosons* and the $(\pi^\pm, \pi^0)$ *bosons* are particle-antiparticle symmetric, and thus each particle state contains a matching particle-antiparticle pair of substates, which we denote here as *energy quarks* (similar to the proton energy quarks E_{q_p} defined in Eq. (3.6), since their quark energies are the main property of interest. Specifically, the *average-energy gauge boson* $\overline{WZ}$ contains the energy-quark pair $E_{gb}\bar{E}_{gb}$, where each quark has the energy value

$$E_{gb} = \frac{E_{\overline{WZ}}}{2} = 42.893\,\text{GeV}. \tag{3.10}$$

Hence, $\overline{WZ}$ has an energy of $2\,E_{gb}$. The top quark t has an energy of $4\,E_{gb}$, as shown in Eq. (3.4). Furthermore, the top quarks have to be produced in matching $(t, \bar{t})$ pairs in order preserve their quantum numbers, where the (unobserved) $T = (t\bar{t})$ virtual toponium resonance has an energy of 8 E_{gb}. Thus, the E_{gb} energy quantum is acting effectively as a *unit energy* for reproducing not only the $\overline{WZ}$ average gauge boson energy, but also the t quark and the toponium particle state T energies. It is useful to re-label the E_{gb} energy quantum as the gauge boson "energy packet" E_{gb}, which occurs in multiples that supply the LHC particle-state energies.

Combining the above results, we can describe the gauge boson wave-packet excitation process as occurring in two steps:

A direct LHC q_p–q_p quark–quark collision concentrates the beam energy, which then creates a factor-of-137 quantized energy α-boost that generates an $E_{gb}\overline{E}_{gb}$ pair of energy-packets.

This energy α-boost collision mechanism is repeated so as to form an energy reservoir of gauge boson energy packets that is large enough to convert into an energy-quantized particle state.

This α-boost energy-packet scenario is reinforced by the fact that the excitation formalism also occurs in the low-energy *boson* particle states, and, with an additional factor of 3/2, in the *fermion* particle states, where the ground state in these cases is an electron–positron pair. These low-energy e–$\bar{e}$ energy α-boosts create the boson and fermion energy packets E_b and E_f, respectively.

We can relate the energy E_t of the top quark to the energy E_p of the proton by combining Eqs. (3.4) and (3.8). This gives the equation

$$E_t = \frac{4E_p}{3\alpha} = 171.44\,\text{GeV}, \tag{3.11}$$

which agrees with the measured top quark energy of 173.0 MeV to an accuracy of 0.8%. The phenomenological significance of this result is that it uses the fine structure constant α to accurately relate the low-energy proton regime to the high-energy LHC top quark regime, and to demonstrate that in the energy representation of the elementary particle spectrum, these two energy regimes coherently merge together.

3.6   Matching *Gauge-Boson* and *Boson* Energy-Packet Excitations

The low-energy *boson* α-boost energy pattern that occurs for the pseudo scalar mesons is strikingly similar to the high-energy *gauge boson* α-boost energy pattern occurs for the high-energy gauge bosons. To illustrate this fact, we plot their energy spectra together on normalized energy scales, using the α-boost ground-state energies as graphical *unit energies*. These energy patterns are displayed in Fig. 3.5.

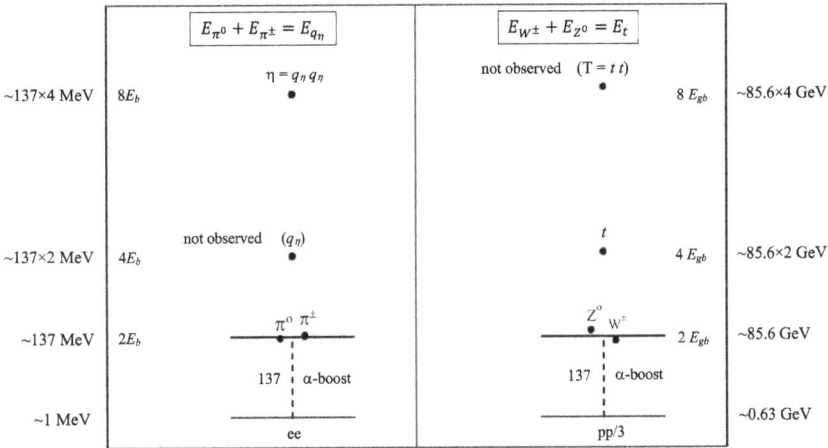

		$E_{\pi^0} + E_{\pi^\pm} = E_{q_\eta}$		$E_{W^\pm} + E_{Z^0} = E_t$	
		$\eta = q_\eta q_\eta$		not observed (T = t t)	
~137×4 MeV	$8E_b$	•		•	$8\,E_{gb}$ ~85.6×4 GeV
		not observed (q_η)		t	
~137×2 MeV	$4E_b$	•		•	$4\,E_{gb}$ ~85.6×2 GeV
		$\pi^0\ \pi^\pm$		$Z^0\ W^\pm$	
~137 MeV	$2E_b$	——•—•——		——•—•——	$2\,E_{gb}$ ~85.6 GeV
		137 ⎸ α-boost		137 ⎸ α-boost	
~1 MeV		————————		————————	~0.63 GeV
		e e		pp/3	

Figure 3.5   Matching excitation-patterns for generating high-energy *gauge boson* and low-energy *boson* particle states. A factor-of-137 α-boost from the pp/3 ground state generates two E_{gb} energy packets at the (W$^\pm$, Z^0) energy level, and a factor-of-137 α-boost from the e$\bar{e}$ ground state generates two E_b energy packets at the (π^0, $\pi^\pm$) energy level. Doubling these energy packets reproduces the t and q_η quark energies, and doubling again reproduces the η and T energies. The q_η and T states (both in parentheses) are not directly observed.

The (W$^\pm$, Z^0) and t particle states are created as excitations of the 0.625 GeV ($q_p + q_p$) ground state (Eq. 3.7), and are plotted on a linear energy scale, with the vertical dotted lines denoting the energies of the E_{gb} gauge boson energy packet combinations. The matching pseudoscalar ($\pi^\pm$, π^0) particle states are created as excitations of the 1.022 MeV (e^-, e^+) ground state, and are similarly plotted, with the vertical dotted lines denoting the energies of the E_b boson energy packet combinations. The top energy pattern is for the *highest* energy particles ever produced, and the bottom energy pattern is for the *lowest*, and yet the two energy patterns are almost identical. The energy scales for the top and bottom excitation patterns differ by a factor of about 630, but the particle excitation processes are the same: repeated α-boosts that create quantized energy packets, which combine together to produce the particle-state energies.

The η meson, which has matching quark and antiquark substates, is displayed in Fig. 3.5 in the form $\eta = q_\eta \cdot \bar{q}_\eta$, where the quark q_η is an "energy quark" whose essential property is its energy E_{q_η}. q_η does not correspond to an observed state, and hence is enclosed in parentheses. It is included in Fig. 3.5 to demonstrate that its quark energy $E_{q_\eta} = 4E_b$ is analogous to the observed t quark energy $E_t = 4E_{gb}$. The $T = t\bar{t}$ toponium state that is indicated in Fig. 3.5 at an energy of $8E_{qb}$ is not observed, and hence is also enclosed in parentheses. It is analogous to the $8E_b$ energy of the observed η meson.

3.7   Boson, Fermion and Gauge Boson Energy Excitation Channels

Three different α-boost energy channels, each with an associated energy packet E and matching antipacket $\overline{E}$ are required empirically in order to reproduce the basic quark and particle states [1]. These are the integer-spin *gauge boson* and *boson* channels (see Fig. 3.1) and the half-integer-spin *fermion* channel. Their energy packets are defined as follows:

$$E_{gb} \equiv E_{\text{gauge boson}} = \frac{E_{q_p}}{\alpha} = \frac{E_p}{3\alpha} = 42.859\,\text{GeV}; \quad \overline{E}_{gb} = \frac{E_p}{3\alpha}. \quad (3.12a)$$

$$E_b \equiv E_{\text{boson}} = \frac{E_e}{\alpha} = 70.025\,\text{MeV}; \quad\quad\quad \overline{E}_b = \frac{E_{\bar{e}}}{\alpha}. \quad (3.12b)$$

$$E_f \equiv E_{\text{fermion}} = \frac{3E_e}{2\alpha} = 105.038\,\text{MeV}; \quad\quad \overline{E}_f = \frac{3E_{\bar{e}}}{2\alpha}. \quad (3.12c)$$

The E and $\overline{E}$ energy packets are created in particle–antiparticle matching pairs via α-boosts in energy from proton quark-pair (3.12a) and electron-pair (3.12b,c) ground states, respectively, in a manner that obeys the over-all electric charge, spin angular momentum, and color charge conservation laws. This E and $\overline{E}$ labeling is also needed because the energy-packet fits to the experimental data indicate that paired quark–antiquark states have a small (strong-interaction) *hadronic binding energy* (HBE) of 2–3% for particle energies of about 1 GeV and below, which decreases with increasing particle energy above 1 GeV, and essentially vanishes at energies of 6 GeV and above (asymptotic freedom). Unpaired states do not show this HBE effect. Applying a systematic empirically-determined HBE correction (which is *not* applied here) would enable us to employ the energy-packet formalism and empirically obtain calculated paired-quark particle energies at a 1% accuracy level, as shown in Tables 3.1 and 3.2.

The *boson* and *fermion* particle states are produced by energy α-boosts from (e^-, e^+) electron-pair ground states. In the *boson* excitation channel, the $\pi = E_b\overline{E}_b$ pion is the lowest particle excitation state, and its E_b and $\overline{E}_b$ energy packets occur in the form of a hadronically-bound $E_b\overline{E}_b$ pair. However, in the *fermion* excitation channel, the $\mu\bar{\mu} = E_f\overline{E}_f$ muon pair is the lowest excitation state, and its E_f and $\overline{E}_f$ energy packets do *not* bind together hadronically. Thus, the generation of fermion particles and antiparticles can be studied as separate (but matching) energy channels. The fermion excitation patterns are displayed in Fig. 3.2 and Table 3.1, and the boson pseudoscalar meson excitation pattern is displayed in Table 3.2.

The fermion energy packet E_f is the only one that corresponds to the energy of an observed particle state: specifically, adding the 0.511 MeV ground-state energy to the 105.549 MeV muon energy gives a total energy of 105.549 MeV, which matches the experimental 105.658 MeV muon energy to an accuracy of 0.10%. This suggests that in the calculation of particle

Table 3.1 Particle-to-electron and quark-to-electron energy ratios.

	Equation	Calc. E	Exper. E	Accuracy
	Particle energy (MeV)			
Muon	$E_\mu = (3/2\alpha + 1)E_e$	105.55	105.66	−0.10%
Proton	$E_p = (27/2\alpha + 1)E_e$	945.9	938.3	0.81%
Tauon	$E_\tau = (51/2\alpha + 1)E_e$	1786.1	1776.8	0.52%
	Constituent-quark energy (MeV)			
u, d	$E_{u,d} = (9/2\alpha + 1)E_e$	316	$313 = E_p/3$	1.0%
s	$E_s = (15/2\alpha + 1)E_e$	526	$510 = E_\phi/2$	3.1%[†]
c	$E_c = (45/2\alpha + 1)E_e$	1576	$1548 = E_{\psi_{1s}}$	1.8%[†]
b	$E_b = (135/2\alpha + 1)E_e$	4727	$4730 = E_{\gamma_{1s}}$	−0.06%
$b + c$	$E_{b+c} = (180/2\alpha + 1)E_e$	6303	$6276 = E_{B_c}$	0.43%
	Constituent-quark energy (GeV)			
t	$E^t = (36/2\alpha^2 + 1)E_e$	169.01	173.21	−0.24%

[†]Small HBE corrections are required for the $s\bar{s} = \phi$ and $c\bar{c} = J/\psi$ mesons, which have energies of less than 6 GeV (see text).

Table 3.2 Pseudoscalar meson particle-to-electron energy ratios.

	Equation	Particle energy (MeV)		
		Calc. E	Exper. E	Accuracy
π	$E_\pi = (2/\alpha + 2)E_e$	141.07	137.27[†]	2.8%[‡]
η	$E_\eta = (8/\alpha + 2)E_e$	561.42	547.86	2.5%[‡]
η'	$E_{\eta'} = (14/\alpha + 2)E_e$	981.38	957.78	2.5%[‡]
K	$E_K = (7/\alpha + 1)E_e$	490.69	495.65[†]	−1.0%

[†]Average particle pair energy
[‡]Small HBE corrections are needed

energies from electron ground states, particle-antiparticle excitations must be used. The total available particle energy is the particle energy plus the electron ground-state energy, which is the formulation used for the energy values displayed in Tables 3.1 and 3.2. In this formulation, the (e^-, e^+) ground-state pair annihilate, and this energy is added to the α-boost excitation energy.

The fermion energy states displayed in Table 3.1 are reproduced as *odd* multiples of E_f energy packets. The multiplication of 105 MeV E_f energy packets to reproduce particle and quark energies occurs in two distinctive excitation patterns. The first is an *energy-packet excitation-doubling* pattern: 2 packets added to the $E_e + E_f$ muon ground-state reproduces the (u, d) constituent-quark energy; 4 packets reproduces the $s$constituent-quark energy; 8 packets reproduces the proton energy; and 16 packets reproduces the tau lepton energy. These quark and particle excitations interleave and form the following excitation-doubling energy sequence:

$$(E_e + E_f) + nE_f \ (n = 0, 2, 4, 8, 16) \rightarrow \mu, (u, d), s, p, \tau. \qquad (3.13)$$

This is the equation that is used for the calculated energy values of these quarks and particles in Table 3.1. Quark and particle energies are reproduced as odd multiples of the fermion energy packet $E_f = 105$ MeV of Eq. (3.13). The μ, (u, d) s, p, τ excitation-doubling states of Eq. (3.13) are shown in a row along the bottom, and the energy-tripling quark sequence s–c–b of Eq. (3.14a) is displayed in the middle. The energy α-boost from the proton (u, d) quarks to the high-energy W, Z, t particle states that is defined in Eqs. (3.4)–(3.10) is also included here. The accurately-calculated energy values that are obtained for these states are summarized in Table 3.1. The excitation pattern only emerges if the muon, (u, d) quarks, strange quark s, proton p, and tauon t are all included in a common energy-generation scheme.

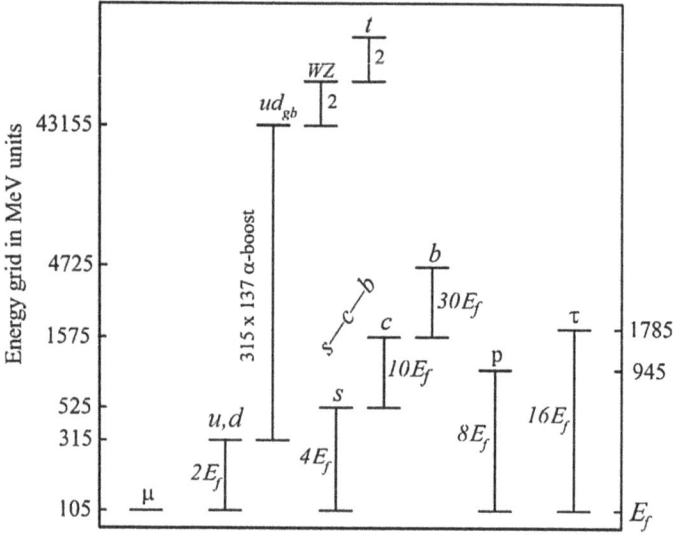

Figure 3.6 Quark and particle energies are reproduced as multiples of the fermion energy packet $E_f = 105$ MeV. The μ, (u, d) s, p, τ excitation-doubling state of Eq. (3.13) are shown in a row along the bottom, and the energy-tripling quark sequence s-c-b of Eq. (3.14a) is displayed in the middle. The energy α-boost from the proton (u, d) quarks to the high-energy W, Z, t particles states is also included here. The accurately-calculated energy values that are obtained for these states are summarized in Table 3.1.

The second fermion energy pattern is a *quark energy-tripling* pattern. It is based on the s quark energy, which is successively tripled and tripled again to create the c and b quarks, which bind together in matching quark–antiquark pairs to form the following vector meson sequence: where

$$E_n = 5nE_f, \quad (n = 1, 3, 9) \Rightarrow s, c, b, \tag{3.14a}$$

$$\text{where } s = 5E_f, \quad c = sss, \quad b = ccc, \tag{3.14b}$$

$$\text{and } \phi = s\bar{s}, \quad J/\psi_{1S} = c\bar{c}, \quad \Upsilon_{1S} = b\bar{b}. \tag{3.14c}$$

Since the s, c and b quarks in Eq. (3.14a) are particle substates that cannot be observed individually, the experimental energies shown for them in Table 3.1 are deduced from the ground-state energies of the vector mesons in Eq. (3.14c) that contain them in matching quark–antiquark pairs. However, the deduced values for the *low-energy* s and c quarks require small HBE corrections, which we do *not* apply.

The effective experimental energy that is cited in Table 3.1 for the $E_{q_p} = E(u,d)$ quarks is $1/3$ of the proton energy, as discussed in Ref. [5]. The theoretical energy equation for the top quark t in Table 3.1 is obtained from an excitation energy path that leads from the electron to the top quark.

Independent confirmation for the constituent-quark energy values of the u, d and s quarks displayed in Table 3.1 is provided by the experimental hyperon magnetic moments, as described in App. D [7]. These results substantiate the values for the excitation-doubled (u, d) and s quark energies that are calculated in Eq. (3.14).

The *boson* π, η, η', K pseudoscalar meson energies are reproduced as multiples of 70 MeV energy packets E_b, as summarized in Table 3.2. The excitation-doubling energy pattern of Eq. (3.14) also occurs here for the *non-strange* π, η and η' mesons, but in an abbreviated form:

$$E_n = (E_e \overline{E}_e + E_b \overline{E}_b) + n E_b \overline{E}_b, \quad \text{where } (n = 0, 3, 6) \to \pi, \eta, \eta'. \quad (3.15)$$

Each of the *strange* K and $\overline{\text{K}}$ mesons has half the energy of the η' mesons: $E_{\text{K}} = 7E_b$. Thus, from an energy standpoint, we have $\eta' = \text{K}\overline{\text{K}}$.

A comparison of the calculated and experimental energies in Table 3.2 reveals that the $E_b \overline{E}_b$ paired-packet π, η, η' mesons require a 2–3% hadronic binding energy correction (which, as we stated above in Table 3.1, is *not* applied), whereas the unpaired-packet K mesons do not require this HBE correction. This is in line with the energy calculations displayed in Table 3.1, where the unpaired-packet μ, p and τ particles do not require HBE corrections.

A ramification of these results is that the spin 0 K mesons have an odd number of E_b energy packets, which implies that the E_b energy packet must have zero spin. This is in agreement with the relativistic spinning sphere

(RSS) equations of App. C [8], where the spin 0 and spin 1/2 E_b and E_f energy spheres are calculated to differ in energy by a factor of $3/2$. This energy factor of $3/2$ is reflected in the *boson* and *fermion* energy-packet values that are displayed in α-boost equations (3.12b) and (3.12c), respectively.

3.8   Energy Paths from the Electron to the t and b Quarks

The elementary particle ground-state energy values indicate that these particle states are formed as combinations of α-quantized energy packets. These packets are created by factor-of-137 α-boosts in energy from well-defined electron and nucleon-quark energy-packet ground states, as defined in Eqs. (3.12a), (3.12b), and (3.12c). These three excitation channels generate unique energy packets — E_{gb}, E_b, E_f, respectively. Each of the three types of energy packet is multiplied by successive ground-state α-boosts, and the packets are stored together so as to create an energy packet reservoir. The stored packets then merge together and reproduce a characteristic sequence of quantized quark and particle energy states, as defined in Eqs. (3.8), (3.11), (3.14–3.16). These three energy reservoirs each generate an *energy stream* that is composed of *energy paths*. The energy streams interlock and form a coherent α-boosted array of energy paths, thus creating an energy generation pattern that covers the whole range of ground-state particle energies. To illustrate this energy pattern, we trace out the path that goes from the electron e to the top quark t (Fig. 3.3), and also the path that goes from the electron e to the bottom quark b (Fig. 3.4).

The energy path from the electron to the top quark can be written out schematically as follows:

$$E_e \Rightarrow E_f \to 3E_f \Rightarrow E_{gb} \to 4E_{gb} = E_W + E_Z = E_t; \tag{3.16a}$$

$$E_e \times \left(\frac{3}{2\alpha}\right) \times 3 \times \left(\frac{1}{\alpha}\right) \times 4 = E_e \times \left(\frac{18}{\alpha^2}\right) = E_{\text{top}}. \tag{3.16b}$$

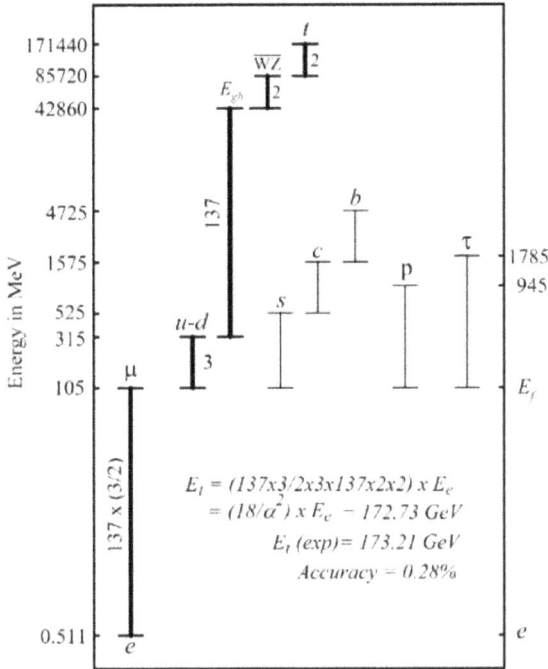

Figure 3.7 The excitation path from the electron to the top quark, which features two factor-of-137 α-boost steps, *fermion* (Eq. (3.12c)) and *gauge boson* (Eq. (3.12a)), that generate quantized energy packets, plus three energy steps where energy packets have merged together. The accuracy of the calculated top quark energy (0.28%) indicates that these sequential energy steps combine together to form a coherent energy path.

It begins with an electron, which is successively excited to form a fermion energy packet (E_f) that is manifested as the leptonic muon, a proton constituent quark $(u\text{–}d)$, a gauge boson energy packet (E_{gb}), and the top quark t. The energy steps along this path consist of factor-of-$(1/\alpha)$ energy boosts (labeled by the symbol $\Rightarrow$) that create the E_f and E_{gb} energy packets, which are followed by packet multiplication steps (labeled by the symbol $\rightarrow$) that denote where energy packets have been merged together to create the energies required for quark and particle states. The sizes of the energy-path steps are defined in Eq. (3.16b). These energy steps are multiplied together to give the following top quark energy equation:

$$E_{\text{top}} = E_e \times \left(\frac{18}{\alpha^2} \right) = 172.73 \,\text{GeV}. \tag{3.17}$$

The currently-listed experimental top quark energy value is [1]

$$E_{\text{top}} = 173.0 \,\text{GeV}. \tag{3.18}$$

The experimental accuracy is $173.0/172.73 = 1.6\%$.

References

[1] M. Tanabashi *et al.* (Particle Data Group) *Phys. Rev. D* **98**, 030001 (2018) (URL: http://pdg.lbl.gov).

[2] Ref. [1], Table 1.1.

[3] D. Hanneke, S. Fogwell, and G. Gabrielse, *Phys. Rev. Lett.* **100**, 1207801 (2008).

[4] A. I. Miller, *137, Jung, Pauli, and the Pursuit of a Scientific Obsession* (Norton, New York, 2009) p. 248.

[5] L. Lederman with D. Teresi, *The God Particle* (Delta Books, New York, 1993), pp. 28–29.

[6] R. P. Feynman, *QED: The Strange Theory of Light and Matter* (Princeton University Press, 1985), p. 129.

[7] R. A. Arndt and M. H. Mac Gregor, "Nucleon-Nucleon Phase Shift Analyses by Chi-Squared Minimization," in *Methods in Computational Physics, Vol. 6. Nuclear Physics.*, B. Alder, S. Fernbach and M. Rotenberg, eds. (Academic Press, New York, 1966), pp. 253–296.

[8] J. D. Jackson, *The Physics of Elementary Particles* (Princeton University Press, 1958), pp. 57–58.

Chapter 4

The Relativistically-Spinning Sphere (RSS), Magnetic Size, and Magnetic Self-Energy of the Electron

4.1 The Enigmatic Electron

The first known elementary particle, the electron, was identified by J. J. Thomson in 1897, over a century ago. As the lowest-mass and stable particle, it should logically be one of the most fundamental, and hence simplest, to interpret. And yet we have not been able to reconcile the salient facts we know about it, which are the following:

1. The electron contains a negative electric charge $-e$. The magnitude of $|e|$ of this charge is precisely the same as that of the $+e$ charge on the proton. The magnitude is also precisely the same for the $+e$ and $-e$ charges on all of the other elementary particles in the universe, and the total electric charge in a closed system is conserved. Thus, the charge e is a physical entity that is not unique to the electron.

2. The electron is in itself more than just its electric charge, but it must have a non-electromagnetic (mechanical) mass component m_e that holds the electron together, since no collection of purely electric charges is stable (Earnshaw's Theorem).

3. The electron is a *lepton*, as are the weakly-interacting muon and tauon. Each of the lepton types has an associated neutrino, and the lepton and

its neutrino each carry the conserved *lepton quantum number* L. The
electron e and electron neutrino ν_e have an *electron* lepton quantum
number $L_e = +1$. The *positron*, the antiparticle of the electron, has
properties that mirror those of the electron, and the positron and its
associated antineutrino each have $L_e = -1$. The total electron lepton
number L_e is a conserved quantity in electron interactions.

4. In electron–electron (Møller) scattering and positron–electron (Bhabha)
   scattering, the total cross section and the angular distribution of the
   scattering are accurately calculated as the interaction of two *point-like*
   charges, which means that they have no detectable size effects, down to
   a length of at least 10^{-16} cm [1]. This information tells us four things:

   a. The charge e on the electron or positron is essentially point-like.

   b. A classical self-interacting spherically-distributed electric charge e
      with the *classical electron* radius $r_e = 2.82 \times 10^{-13}$ cm has the cal-
      culated Coulomb self-energy $E_C = 0.511$ MeV, which is equal to the
      total energy of the electron. The self-energy E_C varies inversely with
      the radius [2]. Thus, a point-like self-interacting charge e with a
      radius of 10^{-16} cm would have an enormous energy. Hence, the point
      charge e on the electron must be a *non*-self-interacting $E_C = 0$ entity.

   c. Since the electric charge on the electron has zero Coulomb self-energy,
      the non-electromagnetic mechanical mass m_e must carry almost the
      entire inertial mass of the electron, namely, $m_e c^2 = 0.511$ MeV with
      an exception for a small dipole magnetic moment energy component.

   d. The calculated Møller and Bhabha scattering cross sections appear to
      be purely electromagnetic, so that the mechanical mass m_e must itself
      be essentially a *non-interacting* entity as compared to the strength of
      the Coulomb interactions.

These problems with the electron have been well-discussed in the scientific literature. In an early review article on the quantum theory of

radiation, Enrico Fermi (1932) made the following summary:

"In conclusion, we may therefore say that practically all the problems in radiation theory which do not involve the structure of the electron have their satisfactory explanation, while the problems connected with the internal properties of the electron are still very far from their solution" [3].

Nearly sixty years later, Asim Barut (1991) restated these difficulties in more detail:

"If a spinning particle is not quite a point particle, nor a solid 3-dimensional top, what can it be? What is the structure which can appear under probing with electromagnetic fields as a point charge, yet as far as spin and wave properties are concerned exhibits a size of the order of the Compton wave length?" [4]

The electron is generally considered to be a point-like particle whose properties cannot be understood classically. However, in the chapter we describe a specific model, a Compton-sized relativistically spinning sphere (RSS) with an equatorial point electric charge e, which uniquely accounts for the main spectroscopic features of the electron, including its first-order anomalous magnetic moment and its point-like scattering behavior. Thus, this model hopefully serves as at least a step forward in the task of resolving the enigmatic electron's nature described above.

4.2 The Mass Increase of a Relativistically Spinning Sphere (RSS)

We consider a uniform sphere of matter, initially at rest, with a mass M_0 and radius R. If we now set it into a spinning motion, it will acquire rotational energy in the form of a relativistic mass increase. If the sphere is spinning with an angular velocity ω around the spin axis, then a mass element at a distance r from the axis will have an instantaneous linear velocity $\nu = \omega r$. In order to calculate the total relativistic mass of the spinning sphere, we divide it up mathematically into mass elements which are equidistant from the axis of rotation. The simplest mass element is a ring center on the rotational axis. The mass of the spinning ring is increased relativistically by the factor:

$$M_S = \frac{M_0}{\sqrt{1 - (\omega^2 r^2 / c^2)}}. \tag{4.1}$$

In order to calculate the mass increase of the spinning sphere, we can use cylindrical polar coordinates (r, θ, z) and combine spinning rings into spinning cylinders, as illustrated in Fig. 4.1, where the spin axis is directed along the z coordinate axis.

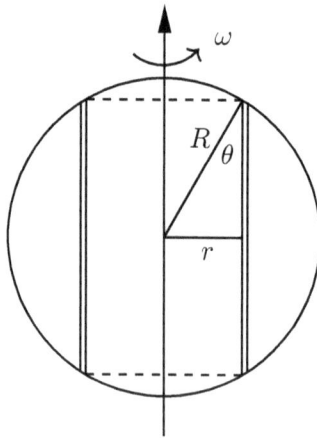

Figure 4.1 A spinning sphere is divided into cylindrical mass elements, where R is the sphere radius, r is the distance from the spin axis, $2R\cos\theta$ is the length of the cylinder, and dr is the wall thickness.

The volume of the cylinder shown in Fig. 4.1 is

$$V(r) = (2\pi r)(2\sqrt{R^2 - r^2})dr = 4\pi R \cos\theta r dr. \tag{4.2}$$

When properly integrated over the radial coordinate, we obtain the volume of the sphere:

$$V = \frac{4\pi R^3}{3}. \tag{4.3}$$

When the electron is spinning at the rotational limit, where the equator of the spin axis is moving at the speed of light (or infinitesimally below this value), the volume of the sphere is given by the equation:

$$V(r) = \frac{V_0(r)}{\sqrt{\left(\dfrac{1 - \omega^2 r^2}{c^2}\right)}}. \tag{4.4}$$

The mass density of the sphere is

$$\rho = \frac{M_0}{V} = \frac{3M_0}{4\pi R^3}. \tag{4.5}$$

When the sphere is set into rotation, it has a relativistic mass M_s. The largest value for angular velocity ω is the value $\omega = c/R$, which represents the angular velocity on the equator of the spinning sphere which is moving at the velocity of light c.

From Eq. (4.5), the relativistic mass of the spinning sphere can be expressed as [5]:

$$M_S = \left(\frac{3}{2}\right) M_0. \tag{4.6}$$

Namely, *the relativistic spinning mass M_S is 3/2 times larger than the non-spinning rest mass M_0.* This is a non-trivial result, and it follows uniquely from the spherical geometry that was selected for the electron.

4.3   The Non-Euclidean Geometry and Mass Density of the RSS

In the above calculations, the geometry of the ring in the rotating frame is not Euclidean; its circumference is greater than $2\pi r$. The calculated volume element inside the rotating sphere is increased by the same relativistic factor that the calculated mass element is increased, so the calculated density remains unchanged.

Theoretically, the topic of relativistically spinning masses has been controversial. The controversy centers around two main points:

1. Is the geometry of the rotating mass non-Euclidean?
2. What is the nature of the purely relativistic stresses, if they exist?

Spinning sphere can be subdivided into spinning discs, which are less complex to deal with mathematically, and are the objects usually discussed. The rotating disc has been extensively analyzed by Henri Arzelies in his book *Relativistic Kinematics*. He comments as follows [6]:

"Despite the divergencies of opinion which appear in the literature, agreement seems to have been reached on the first of these problems, the nature of the spatial geometry on the discs [it is non-Euclidean]. The second problem concerns the stresses of purely relativistic origin which may be created with the discs. It is still the object of numerous arguments among specialists."

The one comment which we can add to this discussion is that if purely relativistic stresses were occurring in a spinning mass element, then it should cause a density perturbation. However, the fact is that the non-Euclidean volume increase and the relativistic mass increase offset one another, and this keeps the density to be constant. Thus, this implies that the purely relativistic stresses are in fact *not* occurring. After all is said, spinning electrons do exist, and their rotational velocity is probably relativistically high.

4.4 The Spin and Magnetic Moment of a Compton-sized RSS

The mass m_e and electric charge $-e$ of the electron can be regarded as fundamental *static* properties, which are more-or-less independent of whatever size we ascribe to them. On the other hand, the spin angular momentum J and magnetic dipole moment μ of the electron are fundamental *dynamical* properties that do depend on size. They are functions of the radius r of the electron, where we assume that the electron mass is in the form of a solid sphere of matter. The spin J is the product of the moment of inertia I and angular velocity ω about the spin axis. The magnetic dipole moment μ is the product of the area A of the current loop formed by a rotating equatorial charge e and the magnitude of the electric current I that corresponds to the circulation of the charge in the loop. In the RSS model, the spin angular moment J has a quadratic dependence on r, whereas the magnetic moment μ has a linear dependence. Thus, a good test of the RSS model is to see if a single value of the radius r can simultaneously fit both the equations for J and μ.

The moment of inertia of a mass element M at a distance r from the spin axis is I $= Mr^2$. This equation can be applied to the RSS relativistic mass integral, and the result becomes

$$I_{rel} = \left(\frac{3}{4}\right) M_0 R^2 = \left(\frac{1}{2}\right) M_S R^2. \tag{4.7}$$

The corresponding non-relativistic moment of inertia is

$$I_{non-rel} = \left(\frac{2}{5}\right) M_0 R^2. \tag{4.8}$$

The larger value for I_{rel} is due to the r^2 weighting of the relativistic mass increase at larger distance r from the axis of rotation.

The maximum allowed value for ω is $\omega = c/R$. The calculated value for J at this limit is

$$J = I\omega = \left(\frac{1}{2}\right) M_S R_c. \tag{4.9}$$

If we now set R equal to the Compton radius $R_C = \hbar/M_S c$, where M_S is the spinning mass, we obtain the spin angular momentum:

$$J = \left(\frac{1}{2}\right) \hbar. \tag{4.10}$$

Equation (4.10) is a *directly calculated quantity*. Thus, the relativistically spinning sphere (RSS) model ties together the mass, spin angular momentum, and Compton radius of the electron. This also provides a reason for the observed spin quantization of electrons; namely, they are all spinning as fast as they are allowed to, with their equators moving at, or infinitesimally below, the velocity of light c.

In a final step, the RSS model of a Compton-sized electron correctly yields both its spin J and magnetic moment μ, and hence its gyromagnetic ratio:

$$g = \left(\frac{\mu}{J}\right)\left(\frac{e}{2mc}\right) = 2. \tag{4.11}$$

This is a classical result that uniquely follows from the systematics of the relativistically spinning sphere model of the electron. We shall study the anomalous magnetic moment of the electron in Chapter 5, because it is important and is the most accurately known quantity in all of physics, both experimentally and theoretically. The RSS model provides an explanation, and it gives an accurate estimate of the sign and magnitude of the anomaly. Thus, this is a strong argument for the realty of a Compton-sized electron.

References

[1] D. Bender *et al.*, *Phys. Rev. D* **30**, 515 (1984), p. 515.

[2] A detailed discussion of the self-energy of the electric charge e of the electron is given in M. H. Mac Gregor, *The Enigmatic Electron, 2nd Edition* (El Mac Books, Santa Cruz, 2013), Ch. 7.

[3] E. Fermi, *Rev. Mod. Phys.* **4**, 87 (1932), p. 87.

[4] A. Barut, in *Fundamental Theories of Physics; V. 45, The Electron: New Theory and Experiment*, D. Hestenes and A. Weingartshofer, eds. Proceedings of the 1990 Antigonish Electron Workshop at St. Francis Xavier University, Antigonish, Nova Scotia (Kluwer Academics, Dordrecht, 1991), p. 109.

[5] M. H. Mac Gregor, *The Energy-Packet Generator 137 and Key Energy Equation $E_W + E_Z = E_{top}$* (El Mac Books, Santa Cruz, 2017), p. 71.

[6] H. Arzelies, *Relativistic Kinematics* (Pergamon Press, Oxford, 1966), p. 235.

Chapter 5   The Answer to Feynman's Challenge

5.1   Richard Feynman's Challenge to Physicists

Two of the most accurately known quantities in physics are (1) the experimental measurement of the anomalous magnetic moment of the electron, and (2) the corresponding theoretical QED calculation of this anomaly. They represent one of the greatest successes in physics, and at the same time one of its greatest mysteries. The experimental measurement and the QED calculation agree to better than ten significant figures, but we have no understanding as to the physical nature of this anomaly. According to the accepted paradigm in physics, the electron is a point particle, which has no measurable size effects down to a radius of 10^{-16} cm or less. Hence its dominant spectroscopic properties — spin J and magnetic moment μ — are beyond the reach of our classical understanding. The Dirac equation supplies values for the spin J and magnetic moment μ. But it has no explanation for the fact that the measured magnetic moment is about 0.1% larger than this value. This situation has been vividly described by Richard Feynman, who in 1961 challenged physicists to come up with an explanation for this anomaly: [1]

> "*It seems that very little physical intuition has yet been developed in this subject. In nearly every case we are reduced to computing exactly the coefficient of some specific term. We have no way to get a general idea of the*

result to be expected. To make my view clear, consider, for
example, the anomalous electron moment.... We have no
physical picture by which we can easily see that the correc-
tion is roughly $\alpha/2\pi$; in fact, we do not even know why
the sign is positive (other than by computing it). ...We
have been computing terms like a blind man exploring a
new room, but soon we must develop some concept of this
room as a whole, and to have some general idea of what is
contained in it. As a specific challenge, is there any method
of computing the anomalous moment of the electron which,
on first rough approximation, gives a fair approximation to
the α term...?"

As we now describe, the Compton-sized RSS relativistic spinning sphere
model provides a unique response to Feynman's Challenge.

5.2   Experimental Results for the Magnetic Moment
Anomaly a

The two main spectroscopic features of the electron are its spin angular
momentum J and magnetic moment μ. In Sec. 4.4, we derived the values for these quantities from the formalism of the RSS spinning sphere.
Theoretically, they are also given by the Dirac equation:

$$J_{\text{Dirac}} = \frac{1}{2}\hbar. \tag{5.1}$$

$$\mu_{\text{Dirac}} = \frac{e\hbar}{2m_e c}. \tag{5.2}$$

In the quantum mechanical spin formalism, these values are the projections
of J and μ along the z axis of quantization, where the electron spin axis itself
is tipped at the angle $\theta_{\text{QM}} = 54.73°$ with respect to the z axis. However,
in the RSS formalism of Sec. 4.4, these are the total values for J and μ,

and the tipping angle θ_{QM} is produced by the requirement of minimizing the electrostatic quadrupole moment of the electron. The RSS formalism can be brought into agreement with the quantum mechanical formalism by increasing its radius by a factor of $\sqrt{3}$, [2] but then we lose some of the relationships embodied in the factor-of-137 radial scaling triplet displayed in Eqs. (3.1) and (3.2) of Chapter 3.

The Dirac gyromagnetic ratio of Eqs. (5.2) and (5.1) is

$$g_{\text{Dirac}} = \frac{\left(\dfrac{\mu_{\text{Dirac}}}{J_{\text{Dirac}}}\right)}{\left(\dfrac{e}{2mc}\right)} = 2. \tag{5.3}$$

However, accurate measurements of g give a value that is about 0.1% larger than this value, so that [3]

$$g_{\text{exper}} = 2(1 + a), \tag{5.4}$$

where a is the dipole magnetic moment anomaly. This anomalous behavior was first noticed (1947) as the *Lamb shift* of atomic levels in hydrogen and deuterium, and its explanation led to the development of the theory of quantum electrodynamic (QED). It can be qualitatively attributed to the fact that Dirac theory does not take into account the (real or virtual) radiative effects that occur when electrons are accelerated.

The most accurate experimental measurements of this anomaly come from the properties of free polarized electrons. If a polarized electron is trapped in an external magnetic field, the magnitude of the anomaly a in Eq. (5.4) can be directly measured [3, 4]. The polarization vector precesses with the angular velocity

$$\omega_{P} = \left(\frac{eB}{mc}\right)\left(\frac{1}{\gamma} + a\right), \tag{5.5}$$

whereas the cyclotron frequency of the electron is

$$\omega_{\text{c}} = \frac{eB}{mc\gamma}. \tag{5.6}$$

The polarization vector changes between longitudinal and transverse at the difference frequency

$$\omega_{\text{D}} = \omega_{\text{P}} - \omega_C = a\left(\frac{eB}{mc}\right). \tag{5.7}$$

Thus, by repeated measurements of the storage time of the electron in the external field and the polarization vector at the time it is ejected, the value of ω_{D}, and hence of the anomaly a, can be accurately determined. This $g-2$ measure yields the value

$$a \cong \frac{\alpha}{2\pi}. \tag{5.8}$$

This is one of the most striking successes of quantum electrodynamics. The challenge posed by Feynman is to provide a physical explanation for this magnetic moment anomaly that gives a good approximation to the value shown in Eq. (5.8).

5.3   The Rasetti and Fermi Magnetic Field Energy Calculation

The electron magnetic moment equation is $\mu = e\hbar/2m_ec$. A key point with respect to this equation is that the magnetic moment μ is a function of just its mass m_e and charge e. This same basic equation also applies to the muon. Both particles have the same charge e, but differ by a factor of 207 in their masses. Taking the ratio of these two equations, we obtain

$$\frac{\mu_{\text{muon}}}{\mu_{\text{electron}}} = \frac{m_{\text{electron}}}{m_{\text{muon}}}, \tag{5.9}$$

which has an experimental accuracy of about 7 parts in a million [5]. This shows that the inverse relationship between μ and m is the same for each of these particles, even though they have radically different masses. Since the magnetic moment μ depends on just the mass m of the particle, the anomalous magnetic moment a can logically also be attributed to m. Thus, we can write

$$\mu_{\text{exp}} = \frac{e\hbar}{2(m - \Delta m)} c. \tag{5.10}$$

As shown in Eq. (5.8), the experimental value for the magnetic moment is

$$\mu_{\text{exp}} \cong \left(\frac{e\hbar}{2mc}\right)\left(1 + \frac{\alpha}{2\pi}\right). \tag{5.11}$$

This equation applies to both the electron and muon, and it gives

$$\Delta m \cong \frac{m \cdot \alpha}{2\pi}. \tag{5.12}$$

Since the anomaly Δm is roughly 0.1% of the mass, we need to identify a mass component within the electron or muon that constitutes 0.1% of the total mass of the particle. There are four kinds of mass (energy) that we can define for an electron or muon:

(*a*) *electrostatic self-energy* W_E;
(*b*) *magnetic self-energy* W_H;
(*c*) *mechanical mass* W_M;
(*d*) *gravitational mass* W_G.

A clue to selecting the relevant mass is provided by the fact that the magnetic moment anomaly also occurs in the gyromagnetic ratio g of Eq. (5.4). Hence this is an anomaly in the magnetic moment, and not in the spin. Viewing the anomaly in the context of Eq. (5.12), it logically corresponds to an *irrotational* mass component that affects μ but not J. The only irrotational mass component in the above list is the magnetic self-energy W_H, which is the magnetic field energy associated with the current loop that generates μ. With respect to the other particle mass-energy components, the mass W_E of the point charge e is zero (Sec. 4.1), and the gravitational mass W_G is negligible at the length scales considered here. This leaves just W_H and W_M as masses that contribute to the gyromagnetic ratio. The mass W_M is responsible for almost all of the inertial properties of the particle, including the spin J and energy mc^2. Hence it is not irrotational. Thus, W_H is singled out as the culprit that is causing the anomaly in g. Since the anomaly is a 0.1% effect, we expect to find $W_H/W_M \cong 0.1\%$. In order to see if the RSS is consistent with this result, we need to calculate W_H.

An interesting historical fact which emerges here is that the key to answering Feynman's 1961 challenge was in essence provided by Franco Rasetti and Enrico Fermi in 1926, 35 years earlier [6]. Fermi was perhaps the first to recognize that these dynamical properties have profound ramifications with respect to the size of the electron. He and Rasetti demonstrated that the magnetic self-energy W_H associated with the magnetic moment μ sets a lower bound on the size of the electron that is considerably larger than the classical electron radius $R_o = e^2/m_ec^2 = 2.8 \times 10^{-13}$ cm. The suggestion was made that the electron *magnetic* size may be much greater than the *electric* charge distribution. However, both the RSS and the Rasetti–Fermi calculation are required in order to establish this result. Rasetti and Fermi's obtained the following estimate of the energy in the magnetic field of the electron:

Using polar cylindrical coordinates (r, θ, z), consider the magnetic moment μ as arising from a current loop whose axis is along the z axis. The asymptotic magnetic field components are [7]

$$H_r = \frac{2\mu \cos\theta}{r^2}, \quad H_\theta = \frac{\mu \cos\theta}{r^2}. \tag{5.13}$$

Assuming that μ is spread uniformly throughout a spherical volume of radius R_H, and that these equations apply all the way back to R_H, integration over the external field gives the following lower bound for the magnetic field energy:

$$W_H^{\text{ext}} = \left(\frac{\mu^2}{8\pi}\right) \int_{R_H}^{\infty} \int_0^{\pi} \left(\frac{1}{r^6}\right)(3\cos^2\theta + 1)2\pi r^2 \sin\theta \, d\theta \, dr$$

$$= \frac{\mu^2}{3R_H^3}, \quad (r > R_H). \tag{5.14}$$

Several years later, Max Born and Erwin Schrödinger [8] made a similar calculation, but now assuming that the magnetic moment was distributed on the surface of the sphere. This gave $W_H^{\text{ext}} = \mu^2/2R_H^3$, in reasonable agreement with the Rasetti–Fermi calculation.

We can extend the Rasetti–Fermi result by making an estimate of the magnetic energy W_H^{int} that is contained inside of R_H. To do this, we evaluate the field Eq. (5.13) at the magnetic radius R_H, and then assume that these values hold constant all the way in to the origin. This gives

$$W_H^{\text{int}} \cong \left(\frac{\mu^2}{8\pi R_H^6}\right) \int_0^{R_H} \int_0^{\pi} (3\cos^2\theta + 1)2\pi r^2 \sin\theta \, d\theta \, dr$$

$$= \frac{\mu^2}{3R_H^3}, \quad (r < R_H). \tag{5.15}$$

Thus, the total magnetic field energy is

$$W_H^{\text{tot}} \cong \frac{2\mu^2}{3R_H^3}. \tag{5.16}$$

In the RSS model (Sec. 4.4), with its equatorial charge distribution, the magnetic field radius R_H is the Compton radius R_C,

$$R_H = R_C = \frac{\hbar}{mc}. \tag{5.17}$$

Substituting Eqs. (5.2) and (5.17) into Eq. (5.16) gives

$$W_H \cong \frac{\alpha}{6} \cdot mc^2. \tag{5.18}$$

Comparing this equation with Eqs. (5.8) and (5.12), we see that

$$\frac{\alpha}{6} \cong \frac{\alpha}{2\pi}, \tag{5.19}$$

so that we have satisfied Feynman's Challenge! Of course, the close agreement shown in Eq. (5.19) must be regarded as somewhat fortuitous, given the approximations used for the magnetic field. But the key points here are that (1) we can unequivocally determine the *sign* of the anomaly (one of Feynman's requirements), and (2) we have supplied the *physical picture* that underlies the anomaly (another Feynman requirement).

5.4   The Magnetic Moment of a Thin Compton-sized Wire

There is one magnetic calculation we have not yet carried out, and which is central to this discussion. This is the calculation of the magnetic self-energy of a current loop. The estimate of the magnetic self-energy W_H of the electron that was made by Rasetti and Fermi (1926) was obtained by taking the known asymptotic fields of a magnetic dipole (Eq. (5.13)) and integrating these asymptotic forms all the way in to an assumed magnetic dipole radius R_H. Having no model for the electron, and being unaware of the (as yet undiscovered) anomaly in the magnetic moment, they had no basis for pinning down the value of R_H, other than to say it must be large enough so that the magnetic self-energy W_H does not exceed the total energy mc^2 of the electron. This line of reasoning leads to a lower limit for R_H that is about $1/10$ the electron Compton radius. Guided by our subsequent knowledge of the magnetic moment anomaly, and also by the arguments developed in Chapter 4 that favor a Compton-sized electron, we proceeded to set $R_H = R_C$, which yielded a Fermi-type estimate for W_H (Eq. (5.18)) that matches the magnitude of the observed magnetic anomaly (Eq. (5.19)). If we now assume that the magnetic moment of the electron arises from a Compton-sized equatorial current loop, which is what we expect from Ampere's hypothesis, we can extend the above discussion by making a direct calculation of the magnetic self-energy that is associated with this current loop.

The calculation of the magnetic field produced by a thin current loop displays some intriguing QED-like characteristics. Let us represent the current loop as a thin wire of radius R_E which is bent into a circle of radius R_C, where $R_C \gg R_E$ is the Compton radius. The self-inductance L of this current loop is [9]

$$L = 4\pi R_C \left\{ \eta \left(\ln 8 \frac{R_C}{R_E} - 2 \right) + \frac{1}{4}\eta' \right\} \equiv 4\pi R_C \cdot B, \qquad (5.20)$$

where η and η' are the permeabilities outside of and inside of the thin wire, and where B denotes the terms in the brackets. The magnetic self-energy W_H of this current loop is

$$W_H = \frac{1}{2}Li^2 = \frac{1}{2}L \left(\frac{e}{2\pi R_C} \right)^2, \qquad (5.21)$$

where

$$i = \frac{e}{2\pi R_C} \qquad (5.22)$$

is the electric current, and where the charge e is assumed to be moving at the velocity of light c. If we now insert the value for L from Eq. (5.20) into Eq. (5.21), we obtain

$$W_H = \left(\frac{e^2}{2\pi R_C} \right) \cdot B = \frac{\alpha}{2\pi} \cdot mc^2 \cdot B. \qquad (5.23)$$

Thus, if we set the bracket term B equal to unity, we obtain the Schwinger magnetic moment correction factor $\alpha/2\pi$ as a direct result of this calculation! Unfortunately, we have as yet no theoretical basis for setting $B = 1$, but it is nevertheless intriguing to see this very complicated QED result emerge from such a simple calculation, especially since this calculation is based on an essentially classical representation of the physical phenomenon that is involved.

If this were all there was to the story, it might be dismissed as just an interesting coincidence. But there is more. Let us now examine the bracket B. The factor R_E that appears in the denominator of the logarithm in Eq. (5.20) is the radius of the electric charge e, which we know experimentally to have the value $R_E < 10^{-16}\,\text{cm}$. [10]. Thus, it is much smaller than the Compton radius $R_C \cong 4 \times 10^{-11}\,\text{cm}$ that appears in the numerator of the logarithm. Hence, this divergent logarithmic term is dominant, and we see from Eqs. (5.20) and (5.23) that

$$W_H \sim \alpha \cdot mc^2 \cdot \ln \frac{R_C}{R_E}. \tag{5.24}$$

This equation is of precisely the same form as the logarithmic singularity that appears in the QED calculation of the electromagnetic self-energy of a particle, as is illustrated for example in Jackson's 1962 book [11]. This indicates that the logarithmic QED divergence which occurs in the electromagnetic self-energy of a particle is a *magnetic* divergence, and not the *electrostatic* divergence that is commonly assumed. It indeed arises as a consequence of the small spatial size of the electric charge e, but it is manifested by its effect on the magnetic field of the electron rather than as an electrostatic self-interaction. In detail, the point-like rotating charge e leads to a large magnetic field in its vicinity, with a consequently large value for the magnetic self-energy W_H, but the charge e does not interact with itself ($W_E = 0$).

There is another aspect to this story about the magnetic self-energy of the electron current loop. Since we know the value for W_H, as deduced from the magnetic moment anomaly, we can work backwards and see what this implies for the quantities that are contained in the bracket B in Eq. (5.20). In particular, we want to see what we can find out about the radius R_E, which is the one truly unknown quantity that is involved. We start by setting $B = 1$, which gives the correct value for the magnetic moment anomaly (Eq. (5.23)). Then we set $\eta = 1$ (the permeability of free space), and we set $\eta' = 0$ (since there is in fact no wire). With these assumptions, the expression for B reduces to

$$\left(\ln 8 \frac{R_C}{R_E} - 2 \right) = 1, \tag{5.25}$$

which yields the following "effective charge radius":

$$R_E^{\text{eff}} \sim \frac{R_C}{2.5} = 1.5 \times 10^{-11} \text{ cm.} \tag{5.26}$$

We thus have a paradox. Starting with a point-like value for the charge R_E (10^{-16} cm or less), we used the assumption that $R_C \gg R_E$ to derive Eq. (5.24). Applying this equation, we obtained an expression for the self-energy W_H of the current loop (Eq. (5.24)). Then, by matching W_H to the magnetic moment anomaly (Eq. (5.23)), we found that a much larger "effective value" for R_E is required in Eq. (5.26) in order to have $B = 1$. But this is in fact what we should expect from QED. The factor that seems to be operating here is the QED phenomenon of *vacuum polarization*. When a point-like electric charge is placed at a certain position in space, it polarizes the surrounding vacuum state, thereby masking some of the effect of the charge. Studies of this vacuum polarization effect [12] indicate that it extends over a distance which is comparable to the Compton radius R_C. This is just the result that we see in Eq. (5.26), where the effective value for R_E exhibits a Compton-like size. Thus, several facets of QED have emerged from this semi-classical current loop calculation: the Schwinger

$\alpha/2\pi$ correction term; the characteristic QED logarithmic divergence; and a manifestation of the effect of vacuum polarization. In terms of the production of magnetic fields, vacuum polarization may be operating here in the form of a magnetic saturation effect that occurs in the immediate vicinity of the electric charge e.

Perhaps the most important fact about this thin current loop with respect to our present discussions is that it is *Compton-sized*. The electron may scatter off other electrons in a point-like manner, but its key dynamical features — its Dirac dipole magnetic moment, anomalous magnetic moment, and gyromagnetic ratio — are all Compton-sized, and these quantities are uniquely reproduced within the context of the RSS relativistically spinning sphere model. It is also worth noting that the circumference of this Compton-sized RSS current loop, whose electric point charge is moving at the velocity c, is one de Broglie wavelength, $\lambda_{\text{de Broglie}} = h \, / \, m_e c$, where h is Planck's constant.

5.5   The QED Calculation of the Anomalous Magnetic Moment a

Quantum electrodynamics (QED) is the theory for calculating the interactions of photons with matter. It is one of the most successful formalisms in physics, and the calculation of the gyromagnetic ratio g of the electron is one of its most famous examples. The calculations in QED are carried out with the application of Feynman diagrams. Each physical quantity that is to be determined involves a set of diagrams, with each diagram portraying a specific interaction. The interaction amplitude of a diagram is obtained from the rules governing the components of the diagram, and the set of diagram amplitudes is summed at the end of the calculation [13, 14]. The key fact here is the manner in which the fine structure constant $\alpha = e^2 / \, \hbar c \cong 1/137$ dominates these calculations. It is convenient to set

$\hbar = c = 1$, which gives $e = \sqrt{\alpha} \cong 0.0854$. The charge e is the coupling constant at the vertex between an electron and a photon in a Feynman diagram.

Figure 5.1 shows arrows that denote an electron which travels from A to B and couples to a photon at the vertex in the diagram. This diagram represents the generation of the Dirac magnetic moment of the electron. It displays the simplest interaction between an electron and a photon. The magnitude of the amplitude is dictated by the coupling factor $e \cong 0.0854$ at the vertex of the interaction.

Figure 5.2 represents the first-order correction term to the Feynman diagram of Fig. 5.1. It is a "one-loop" diagram in which the electron takes a more complex path going from A to B. Starting at A, it emits a virtual photon, then couples to the magnetic moment field at the vertex, and finally reabsorbs the virtual photon before reaching B. The emission and absorption vertexes each contribute one additional factor of e to the amplitude, so that the amplitude of this diagram involves two extra couplings, and is a factor of order $e^2 = \alpha$ smaller than the amplitude in Fig. 5.1. The detailed calculation of this correction term was made independently by Schwinger, Feynman and Tomonaga, and gave the answer that the *two-extra-coupling* amplitude for the anomalous correction factor is $\alpha/2\pi$. The next step in the QED calculation involves *four* extra couplings, and took three years to

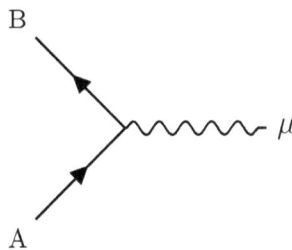

Figure 5.1 The Feynman diagram for the Dirac electron magnetic moment, $\mu_{\text{Dirac}} = e\hbar/2m_e c$.

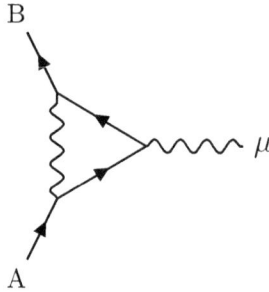

Figure 5.2 The Feynman diagram for the Schwinger factor $a \cong \alpha/2\pi$, where a is the anomalous magnetic moment of the electron, so that $\mu_{\text{electron}} = \mu_{\text{Dirac}}(1 + a)$.

correctly compute. The third step, with *six* extra couplings, took twenty years to complete [14]. The experimental determination of the anomalous magnetic moment was carried out by the precessional gyromagnetic $g - 2$ measurements described in Sec. 5.2.

In order to demonstrate the essential role played by the fine structure constant α in the evaluation of the anomalous magnetic moment a, it is useful to display the results as expansions of the electron gyromagnetic ratio g in powers of α [3].

$$g = 2(1 + a), \tag{5.4}$$

$$a_{\text{QED}} = 0.5 \left(\frac{\alpha}{\pi}\right) - 0.32848 \left(\frac{\alpha}{\pi}\right)^2 + (1.49 \pm 0.2) \left(\frac{\alpha}{\pi}\right)^3. \tag{5.27}$$

$$a_{\text{exp}} = 0.5 \left(\frac{\alpha}{\pi}\right) - 0.32848 \left(\frac{\alpha}{\pi}\right)^2 + (1.68 \pm 0.33) \left(\frac{\alpha}{\pi}\right)^3. \tag{5.28}$$

As can be seen, the coupling constant α is dominant for all orders of the Feynman diagram expansion.

Finally, we display the latest calculated and experimental values for the electron anomalous magnetic dipole moment a [15].

$$a_{\text{QED}} = 0.001\ 159\ 652\ 181\ 643(764). \tag{5.29}$$

$$a_{\text{exp}} = 0.001\ 159\ 652\ 180\ 73(28). \tag{5.30}$$

The close agr eement between the QED calculation of a and the experimental value stands as the most accurately verified prediction in the history of physics [15].

We can add one additional interpretative comment. The second-order magnetic moment Feynman diagram in Fig. 5.2, which is the one that gives rise to the Schwinger $\alpha/2\pi$ correction term, corresponds to the process in which a single virtual photon is emitted and subsequently reabsorbed while the electron is interacting with an external magnetic field. In a discussion on the interpretation of electron mass renormalization diagrams, Feynman [16] points out that the single-virtual-photon diagrams are suggestive of a correction term to the mass of the electron. Thus, the procedure introduced here of identifying the anomalous magnetic moment of the electron with a mass term in the electron (Eqs. (5.10) and (5.12)) seems to be in accordance with the formalism of quantum electrodynamics.

5.6   Concluding Remarks

Appendix A (particle energies) and Appendix B (particle lifetimes) contain the elementary particle data bases for the present book. They represent more than a century of unparalleled experimental and theoretical research into the nature of the elementary particle. The information they contain is the bed rock for any phenomenological or theoretical interpretations we create in our attempts to explain them. The results we have presented in this book have been guided at every step by the messages they seem to be sending us. In the opinion of the present author, the dominant conclusion which has emerged is that the renormalized fine structure constant $\alpha = e^2/\hbar c \cong 1/137$ plays a key role in particle energy generation, particle decays, and atomic structure. Energy "α-boosts" create quantized energy packets that combine together to form the observed energy patterns of leptons, constituent quarks, and ground-state hadrons. The α-quantized energy substructures of these states lead in turn to the α-quantized quark ground-state lifetime groups. And the extension of the constant α itself to include internal radial (r) and mass/energy (m) degrees of freedom has revealed an energy-to-mass phase transition that converts coulomb energy into electron–positron pairs. The electron–positron pairs then act as ground states for the generation of higher-mass particles. The spectroscopic properties of the electron can be accounted-for mathematically by the Compton-sized relativistically-spinning sphere (RSS) model. And, in response to a challenge to theorists that Richard Feynman made in 1961, the electromagnetic features of the RSS electron model can be used to obtain a first-order estimate of the sign and magnitude of the anomalous magnetic moment of the electron.

In the area of atomic structure, the masses of the proton and electron and the value of the fine structure constant α are sufficient to calculate the radius, velocity and binding energy of an electron in the first Bohr orbit of hydrogen.

The number 137 is an important factor in all of the above results. Its ubiquitous role in elementary particle physics has not yet been sufficiently acknowledged. Also, the use of *energy* rather than *mass* in looking for particle generation patterns has not been generally recognized. It is only the *energy patterns* that tie together quarks, leptons and hadrons in one comprehensive picture. And it is the *energy* relationship $E_{\mathrm{W}} + E_{\mathrm{Z}} = E_{\mathrm{top}}$, the surprise result that the LHC researchers had been looking for, that has tied together the elementary particle systematics in the two observed particle energy ranges — the low energy range up to 12 GeV, and the high energy range above 80 GeV. The LHC α-boost in energy from a head-on quark-quark collision ground state (which occurs only once in 10^{10} proton-proton collisions) to the average W + Z gauge boson energy is accurately a factor of 137.

The results obtained in this book can hopefully be construed as a partial reward for all of the physicists who have been faithfully writing the number 137 on their blackboards, as Richard Feynman suggested many years ago.

References

[1] R. P. Feynman, in *The Quantum Theory of Fields*, Proceedings of the Twelfth Conference on Physics at the University of Brussels, October, 1961, ed. R. Stoops (Interscience, New York), pp. 75–76.

[2] M. H. Mac Gregor, *The Enigmatic Electron* (Kluwer, Dordrecht, 1992), Ch. 14.

[3] J. Kessler, *Polarized Electrons* (Springer-Verlag, Berlin, 1976), Sec. 7.2.

[4] A. A. Schupp, R. W. Pidd and H. R. Crane, "Measurement of the g Factor of Free, High-Energy Electrons," *Phys. Rev.* **121**, 1–17 (1961). As a personal note here, Prof. Crane was a member of the late author's Ph.D. thesis committee at the University of Michigan.

[5] B. E. Lautrup, A. Peterman and E. de Rafael, *Physics Reports* **3C**, 193 (1972).

[6] F. Rasetti and E. Fermi, *Nuovo Cimento* **3**, 226 (1926).

[7] J. D. Jackson, *Classical Electrodynamics* (Wiley, New York, 1962), p. 143.

[8] M. Born and E. Schrödinger, *Nature* (London) **135**, 342 (1935).

[9] W. R. Smythe, *Static and Dynamic Electricity* (McGraw-Hill, New York, 1939), p. 316, Eq. (2).

[10] Ref. [1], p. 127.

[11] J. D. Jackson, *Classical Electrodynamics* (Wiley, New York, 1962), p. 593.

[12] D. I. Blokhintsev, *Space and Time in the Microworld* (Reidel, Dordrecht, 1973).

[13] See "Quantum electrodynamics" in Wikipedia.

[14] Ref. [6], pp. 115–119.

[15] See "Anomalous magnetic dipole moment" in Wikipedia.

[16] R. P. Feynman, *Theory of Fundamental Processes* (Benjamin, New York, 1961), p. 140.

Final Editorial Notes

In the notes of his manuscript for The Alpha Sequence,
*Dr. Malcolm H. Mac Gregor started to provide additional
thoughts on the applications of energy packets, specifically,
to the extension to hybrid particle excitations and to the
recent discovery of pentaquarks. It is left to the inter-
ested reader to develop this study of the fine structure constant's connection to the pentaquarks. Perhaps, there is a
future scientist who will continue the research found in this
book and who will make important discoveries, leading to
a Nobel Prize in physics. We shall see.*

— David Akers

Mac Gregor's Closing Notes:

1   The Extension to Hybrid Excitations and Pentaquarks

The basic particle states considered here each contain only one type of
energy packet, which is produced in multiples so as to create quarks and
particles. However, higher-energy and more complex particle states require
a "hybrid" energy content that combines fermion and boson energy packets. The lowest-energy example is the 775 MeV spin 1 meson pair, which

correspond to a 630 MeV spin 1 *fermion* pair of u and d proton constituent quark energy packets combined with a 140 MeV spin 0 *boson* pair of pion constituent-quark energy packets. As another example, the Λ, $\sum$, Ξ, Ω hyperons, which are excited states of the proton, have energies that correspond to three *fermion* energy quarks combined with several *boson* energy packets. The simplest example in the strange particles is the spin 1 $K^*(892)$ meson, which is composed of a 525 MeV strange meson and three 140 MeV *boson* energy packets.

2   The Extension to Conjugate α-Quantized Particle Lifetimes

Quantum mechanically, energy and time are conjugate variables. As applied to elementary particles, the particle and quark *energies* and *masses* (which are linearly related to one another by the constant factor c^2) and the particle and quark *mean lifetimes* are conjugate variables. Hence if the particle and quark ground-state energy configurations are dominated by the fine structure constant $\alpha \sim 1/137$, as demonstrated here in Chapter 1, their corresponding lifetimes should also exhibit a dependence on α. In Chapter 2, we analyzed the global pattern of particle lifetimes. A lifetime quantization in factors of 137 is clearly in evidence, and it is contained in the quarks themselves. In fact, the experimental quark α quantization is so sharply defined that we can deduce the existence of particle quarks directly from the lifetimes themselves, apart from the evidence provided by the particle energies. When taken together, the energy systematics and lifetime systematics form a consistent α-quantized phenomenological formalism that stands apart from the specific theories which are devised to account for this formalism. The experimental data, which are compiled in Appendix A (energies) and Appendix B (mean lifetimes), tell their own story.

Acknowledgments
by Eleanor Mac Gregor

I wish to extend my appreciation to my husband's colleagues whose conversations and perceptive comments over many years influenced the ideas that eventually took form in this present book. Malcolm regarded them not only as fellow workers, but as very dear friends. They appeared so often in Malcolm's conversations with me, that I too came to regard them as my friends. I am thankful to Paolo Palazzi of CERN, to David Akers of Lockheed Martin Skunk Works, to Professor Manuel de Llano of Instituto de Investigaciones Economicas, to Dr. Peter Cameron of Brookhaven National Laboratory, and to Apeiron publisher Roy Keys. All have expressed deep confidence in my husband's work from the early days of the "enigmatic electron" to his most recent work on the alpha sequence. I want to mention that our daughter Elise Ferrell contributed many hours toward putting the manuscript into book form, and together with her father in his last day arrived at a title for the book, *The Alpha Sequence.*

Now I wish to express my feelings of gratitude to my husband of seventy married years. I am an artist, but in time I became married to his science — to his electron. In their books, physicists often give thanks to their wives for their patience. I have a different story. In the late eighties, when our children were grown and out on their own, the dinner table held just the two of us. Our conversations centered on physics. I listened. I learned about the electron, and then more and more about its relationship to other stable elementary particles: the photon, the proton, and the neutrino. It took a lot of study and time for me to understand his work. The joy it gives me now is worth it.

Postscript
A Profile of an Unconventional Thinker in Physics

This essay is an invitation, extended to members of the physics community and people from other disciplines, to join me in viewing the life and thoughts of Malcolm H. Mac Gregor (1926–2019), a physicist who devoted sixty-one years to researching the electron, particle physics, and nuclear physics.

Malcolm was a true "Renaissance Man." His interests carried him into venues beyond science. An accomplished pianist, he had a wide understanding and appreciation of the visual arts. He enjoyed people. His friendship with other physicists was deep, sincere and even affectionate. But, most of all, his closest friends and family may remember him for his wit. I (his wife) particularly enjoyed conversing with him about physics, and I sorely miss hearing his latest revelations about the electron.

As a ten-year old boy, Malcolm performed in his first piano workshop. At that same age, he earned a certificate as Eagle Scout in the Boy Scouts of America. During high school, he participated in three different sports, and the debate team. He entered the U.S. Navy in 1944, the same year he got his diploma. He then served ten years in the Naval Reserve as Ensign. Upon returning to civilian life after World War II, Malcolm earned a Ph.D. from the University of Michigan, in 1953. He was granted an Atomic Energy Commission pre-doctoral fellowship, 1951–1953. During his college years, he played catcher on the physics department softball team. By then, he had married and began raising his first child. Having completed his formal education, Malcolm took a job at the Livermore Radiation Laboratory (LRL) managed by the University of California.

When Malcolm's family moved to California, his professional life began. In 1958, he served as Rapporteur at the International Conference on Nuclear Physics in Paris, France. Nobel Physics Laureate and Professor Aage Bohr invited him to the Niels Bohr Institute on a NATO Fellowship in 1960. In 1962–1963, he was Lecturer at the University of California, Berkeley, and in 1963–1970, he acted as thesis adviser in theoretical physics at U.C. Berkeley. He organized and served as principal speaker of the 1967 International Conference on Nucleon-Nucleon Scattering in Gainesville, Florida (with Richard Wilson, Harvard and Alex Green, Florida). In 1967, he was dedication speaker for ETH Zurich, Switzerland. He served as Chairman of the Nucleon-Nucleon Conference in Dubna, USSR in1968. About this time, he was offered a Full Professorship and Special Chair at Virginia Polytechnic Institute, and elected a Fellow of the American Physical Society. He gave an invited paper at the Coral Gables 1971 Conference of High Energy Physics.

Malcolm's first major book, *The Nature of the Elementary Particle*, published by Springer Verlag, came out in 1978. Later, in 1992, Kluber published his popular book *The Enigmatic Electron*. In 2007, World Scientific published *The Power of Alpha*. This book focused on the mass generation of elementary particles. It contains physics not included within the Standard Model as it is now formulated, while at the same time, conforming to the major results of the Standard Model. Editorial assistance and comments were provided for the book by physicists Paolo Palazzi and David Akers, in 2006.

Because *The Enigmatic Electron* was still selling in 2013, Malcolm self-published a 2nd edition. Around this time, he was elected into the *Emeriti Association* at the University of California, Santa Cruz, where he attended physics seminars. In 2018, he self-published a book titled *The Energy Packet Generator: 137* on Amazon, which shared his newest ideas.

During the same year, the Indian Journal of Physics published two articles representing Malcolm's new ideas. In reference to Malcolm's second article — on the a-quantization of lepton, quark and particle lifetimes — Subham Majumdar Editor-in-chief of the journal wrote in his email of December 18, 2017: *"This paper presents an update on the extensive analysis of data for particles carried out by the author since the 1960's. I have read the recently published Ref. 2 by the same author and the present paper supplements the analysis published there. I think the subject of these papers is very important from the standpoint of Foundations of Physics. What is considered is the issue of the origin of mass. The interpretation of the reported dependence on alpha of the lifetimes and mass should lead to a better understanding of which kinds of energies are involved in rest energies."*

Malcolm's sixty-one years of research can be characterized by a willingness to deal with new ideas that did not fit perfectly into establishment physics. He was aware that differing with the establishment brought censure, and did not regard himself as being outside the current physics of the Standard Model. His emphasis centered upon evaluating particle "behavior patterns" as numbers based solely on experiment. This numerical approach (data) was interpreted by some critics as phenomenology, which Malcolm believed was a valid approach to understanding elementary particle phenomena. In going over his papers, I found he repeatedly weathered criticism and rejections. Sometimes, he was denied opportunities for resubmission, despite having spent fifty-years, himself, refereeing submissions for *Physical Review* and *Physical Review Letters*.

Between 1958 and 1983, Malcolm gave 17 invited talks around the world. He spoke at invited colloquia about elementary particles at 20 universities in the U.S. and Canada.

Upon Malcolm's passing, his professional associates, who were also his friends, expressed the comments given below.

PAOLO PALAZZI: "In 1973 I found a thick physics paper by Malcolm in the SLAC library; very original viewpoint, very interesting! I was just beginning to work on particle systematics in my spare time as a hobby, and noticed some over-lap with Malcolm's work; from that day, I collected and read all his papers and books that I could find, and through the years I developed a sizable Mac Gregor section in my physics papers collection, and a high respect for an unconventional scientist. I am convinced that the day will come when Malcolm's pioneering work in particle mass and lifetime systematics will be recognized, but most likely this will not happen during my lifetime, although I am almost 20 years Malcolm's junior."

DAVID AKERS: "...his work was based upon experimental data and his ideas over-lapped with my own unorthodox ideas on the existence of Dirac magnetic monopoles in nature. My work in particle physics, specifically, magnetic monopoles occurred in the 1980s up to my last publication in 1994; I made connections to the fine structure constant and to the masses of elementary particles after Malcolm did in earlier years. It was a privilege to edit and to make comments on Malcolm's book *The Power of Alpha* in 2006. We met and corresponded throughout the years, even after his retirement from Lawrence Livermore National Laboratory, and we respected each other's ideas on elementary particle physics."

MANUEL de LLANO was at North Carolina University in Wilmington, giving a talk, when he received my email informing him of Malcolm's passing. He said that the day before hearing from me, he had shown Malcolm's "remarkable table on his alpha constant work".

PETER CAMERON: "He was such an important person in the development of my worldview [in physics]; [he was] absolutely essential."

ROY KEYS: "I always looked forward to emails from Malcolm, as they were invariably models of wit and wisdom. Malcolm's hand may someday steer the ship of physics through uncharted waters and on to momentous new discoveries. I feel confident that his legacy is assured."

Please share the following treasured photos with me.

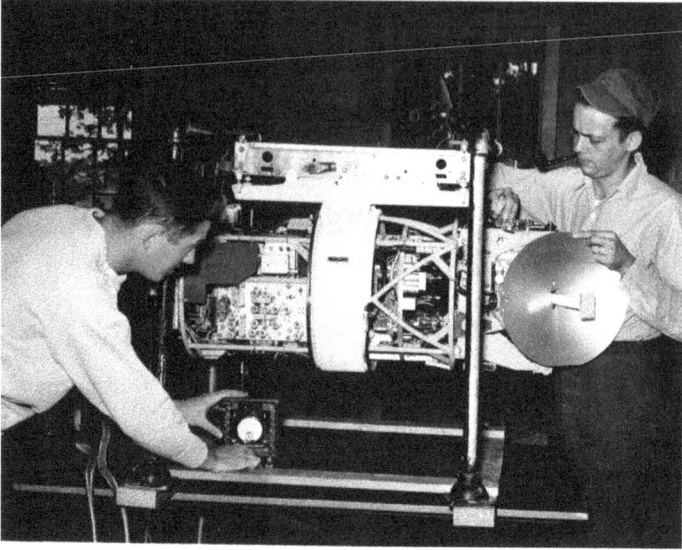

MacGregor working on University of Michigan research project, in high towers, as radio technician, 1948.

Neils Bohr Institute, NATO Fellowship. MacGregor was invited by Aage Bohr, 1960.

Dedication of cyclotron, in Zurich, Switzerland. MacGregor was Keynote Speaker, 1967.

MacGregor as Organizer and Principal Speaker of the International Conference on Nucleon-Nucleon Scattering, (with Richard Wilson, Harvard, and Alex Green, Florida) in Gainesville, Florida, 1967.

Playing at a theatrical event, Walnut Creek, California, 1963.

Playing at a Piano Workshop, in Santa Cruz, California, 2018.

MacGregor enjoying art at the Louvre, Paris, 1969.

MacGregor and wife Eleanor in Paris, 1964.

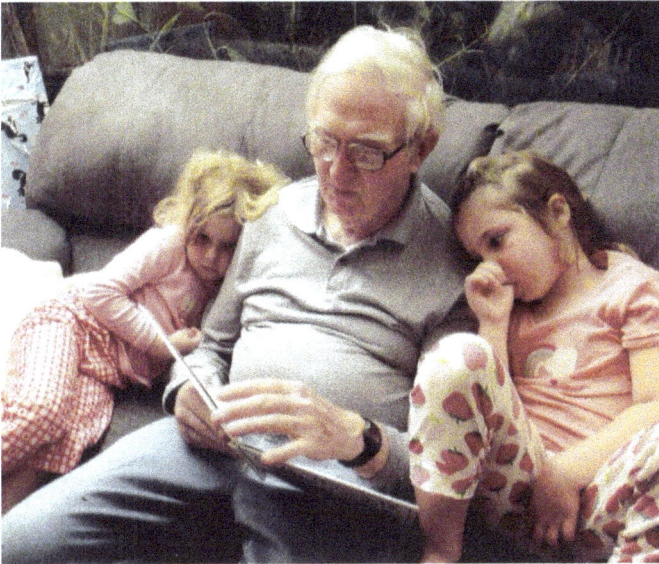

MacGregor reading to granddaughters Ayla and Ariel Ferrell, 2009.

MacGregor and Eleanor windsurfing at Lake Del Valle, California, 1980's.

MacGregor's drawing "My Lady" Vision of his Anima, 2017. Posthumous discovery, in a drawer with his "dream book."

MacGregor's painting at a family event where everyone was asked to draw their own Mandala, 2008.

Appendix A. Review of Particles (2018) Lifetime

Database: 213 Particle Mean Lifetimes

This compilation is from M. Tanabashi *et al.* (Particle Data Group) *Phys. Rev. D* **98**, 030001 (2018) (URL: http://pdg.lbl.gov).

The experimentally-determined mean lifetimes τ for 213 elementary particles are listed here, together with the experimentally-measured full widths $\Gamma = \hbar/\tau$ that are used to calculate the mean lifetimes for the short-lived particles. The two stable particles, the electron and proton, are included in this listing. Lifetimes have not been measured for the Higgs boson and $\Lambda_c \, (2770)^0$ hyperon that are listed in Appendix B.

Mean life	Full width	Particle
τ (sec)	Γ (MeV)	

Long-lived unpaired-quark threshold-state particles

Stable baryon and lepton

1	infinite	proton
2	infinite	electron

Unstable baryon and lepton

3	8.800E+02	neutron
4	2.1969811E-06	muon

u,d,s-quark pseudoscalar mesons

5	5.1160E-08	K^o_L
6	2.6033E-08	$\pi^\pm$
7	1.2380E-08	$K^\pm$

s-quark meson and hyperons

8	2.9000E-10	Ξ^o
9	2.6320E-10	Λ^o
10	1.6390E-10	Ξ^-
11	1.4790E-10	Σ^-
12	8.9540E-11	K^o_S
13	8.2100E-11	Ω^-
14	8.0180E-11	Σ^+

b-quark mesons and baryons

15	1.6380E-12	$B^\pm$
16	1.6400E-12	Ω^-_b
17	1.5710E-12	Ξ^-_b
18	1.5200E-12	B^o
19	1.5090E-12	B^o_s
20	1.4790E-12	Ξ^o_b
21	1.4700E-12	Λ^o_b

c-quark mesons and baryons and τ lepton

22	1.0400E-12	$D^\pm$
23	5.0700E-13	$B^\pm_c$
24	5.0400E-13	$D^\pm_s$
25	4.4200E-13	Ξ^+_c
26	4.1010E-13	D^o
27	2.9030E-13	τ
28	2.0000E-13	Λ^+_c
29	1.1200E-13	Ξ^o_c
30	6.9000E-14	Ω^o_c

Short-lived paired-quark and radiative decay particles

u,d-quark mesons

31	8.5200E-17		π°
32	5.0245E-19	1.3100E-03	η
33	3.3582E-21	1.9600E-01	η'
34	1.5491E-22	4.2490E+00	$\phi(1020)$
35	7.7528E-23	8.4900E+00	$\omega(782)$

s-quark baryon

36	7.4000E-20	Σ°

b-bbar mesons

37	3.2392E-20	2.0320E-02	Υ_{3s}
38	2.0582E-20	3.1980E-02	Υ_{2s}
39	1.2185E-20	5.4020E-02	Υ_{1s}

c-cbar mesons

40	7.0852E-21	9.2900E-02	J/Ψ_{1s}
41	7.8922E-21	8.3400E-02	$D^{*\pm}(2010)$
42	2.2388E-21	2.9400E-01	$\Psi(2s)$
43	2.0698E-23	3.1800E+01	η_{c1S}
44	5.8249E-23	1.1300E+01	η_{c2S}

Top quark and gauge bosons

45	4.6682E-25	1.4100E+03	t
46	3.1569E-25	2.0850E+03	$W^\pm$
47	2.6379E-25	2.4952E+03	Z°

Short-lived excited-state particles

Unflavored mesons

48	4.4534E-24	1.4780E+02	$\rho^\circ(770)$
49	9.4030E-24	7.0000E+01	$f_\circ(980)$

50	8.7762E-24	7.5000E+01	$a_o(980)$
51	1.8284E-24	3.6000E+02	$h_1(1170)$
52	4.6353E-24	1.4200E+02	$b_1(1235)$
53	3.5560E-24	1.8670E+02	$f_2(1270)$
54	2.8996E-23	2.2700E+01	$f_1(1285)$
55	1.1967E-23	5.5000E+01	$\eta(1295)$
56	6.2687E-24	1.0500E+02	$a_2(1320)$
57	1.9946E-24	3.3000E+02	$\pi_1(1400)$
58	1.2881E-23	5.1100E+01	$\eta(1405)$
59	1.1989E-23	5.4900E+01	$f_1(1420)$
60	3.0615E-24	2.1500E+02	$\omega(1420)$
61	2.4838E-24	2.6500E+02	$a_o(1450)$
62	1.6455E-24	4.0000E+02	$\rho(1450)$
63	7.7437E-24	8.5000E+01	$\eta(1475)$
64	6.0386E-24	1.0900E+02	$f_o(1500)$
65	9.0166E-24	7.3000E+01	$f'_2(1525)$
66	2.7312E-24	2.4100E+02	$\pi_1(1600)$
67	3.6365E-24	1.8100E+02	$\eta_2(1645)$
68	2.0896E-24	3.1500E+02	$\omega(1650)$
69	3.9179E-24	1.6800E+02	$\omega_3(1670)$
70	2.5316E-24	2.6000E+02	$\pi_2(1670)$
71	4.3881E-24	1.5000E+02	$\phi(1680)$
72	4.0883E-24	1.6100E+02	$\rho_3(1690)$
73	2.6328E-24	2.5000E+02	$\rho(1700)$
74	4.7353E-24	1.3900E+02	$f_o(1710)$
75	3.1645E-24	2.0800E+02	$\pi(1800)$
76	7.5657E-24	8.7000E+01	$\phi_3(1850)$
77	2.8009E-24	2.3500E+02	$\pi_2(1880)$
78	1.3945E-24	4.7200E+02	$f_2(1950)$
79	3.2585E-24	2.0200E+02	$f_2(2010)$
80	2.5611E-24	2.5700E+02	$a_4(2040)$

81	2.7773E-24	2.3700E+02	$f_4(2050)$
82	7.9300E-24	8.3000E+01	$\phi(2170)$
83	4.4175E-24	1.4900E+02	$f_2(2300)$
84	2.0441E-24	3.2200E+02	$f_2(2340)$

Strange mesons

85	1.3086E-23	5.0300E+01	$K^{*\pm}\,(892)$
86	1.3915E-23	4.7300E+01	$K^{*o}\,(892)$
87	7.3135E-24	9.0000E+01	$K_1(1270)$
88	3.7828E-24	1.7400E+02	$K_1(1400)$
89	2.7890E-24	2.3600E+02	$K^*(1410)$
90	2.4378E-24	2.7000E+02	$K_o^*(1430)$
91	6.6824E-24	9.8500E+01	$K_2^*(1430)^{\pm}$
92	6.0386E-24	1.0900E+02	$K_2^*(1430)^{o}$
93	2.0441E-24	3.2200E+02	$K^*(1680)$
94	3.5388E-24	1.8600E+02	$K_2(1770)$
95	4.1397E-24	1.5900E+02	$K_2^*(1780)$
96	2.4932E-24	2.6400E+02	$K_2(1820)$
97	3.3243E-24	1.9800E+02	$K_4^*(2045)$

Charm mesons

98	2.4652E-24	2.6700E+02	$D_o^*(2400)^{o}$
99	2.1308E-23	3.1700E+01	$D_1(2420)^{o}$
100	1.3857E-23	4.7500E+01	$D_2^*(2460)^{o}$
101	1.4985E-23	4.6700E+01	$D_2^*(2460)^{\pm}$

Bottom mesons

102	2.3935E-23	2.7500E+01	$B_1(5721)^{o}$
103	2.1233E-23	3.1000E+01	$B_1(5721)^{+}$
104	3.2910E-23	2.0000E+01	$B_2^*(5747)^{+}$
105	2.7199E-23	2.4200E+01	$B_2^*(5747)^{o}$

| 106 | 1.0616E-23 | 6.2000E+01 | $B_J(5970)^+$ |
| 107 | 8.1259E-24 | 8.1000E+01 | $B_J(5970)^0$ |

Charm J/Ψ excitations

108	6.0946E-23	1.0800E+01	$\eta_{c0}(1p)$
109	7.8359E-22	8.4000E-01	$\eta_{c1}(1p)$
110	9.4030E-22	7.0000E-01	$h(c)_{1P}$
111	3.3412E-22	1.9700E+00	$\eta_{c2}(1p)$
112	2.4199E-23	2.7200E+01	$\Psi(3770)$
113	2.3258E-23	2.8200E+01	$Z_c(3900)$
114	3.2911E-23	2.0000E+01	$X(3915)$
115	2.7426E-23	2.4000E+01	$\eta_{c2}(3930)$
116	7.3361E-23	1.3000E+01	$X(4020)$
117	8.2276E-24	8.0000E+01	$\Psi(4040)$
118	2.9919E-23	2.2000E+01	$\eta_{c1}(4140)$
119	9.4039E-24	7.0000E+01	$\Psi(4160)$
120	1.1967E-23	5.5000E+01	$\Psi(4260)$
121	1.3432E-23	4.9000E+01	$\eta_{c1}(4274)$
122	6.8573E-24	9.6000E+01	$\Psi(4360)$
123	1.0616E-23	6.2000E+01	$\Psi(4415)$
124	3.6365E-24	1.8100E+02	$Z_c(4430)$
125	9.1418E-25	7.2000E+01	$\Psi(4660)$

Bottom Upsilon excitations

126	3.2108E-23	2.0500E+01	Υ_{4s}
127	1.2906E-23	5.1000E+01	$\Upsilon(10860)$
128	1.3433E-23	4.9000E+01	$\Upsilon(11020)$

N baryon excitations

129	1.8806E-24	3.5000E+02	$N(1440)1/2+$
130	5.4747E-24	1.1000E+02	$N(1520)3/2-$
131	4.3881E-24	1.5000E+02	$N(1535)1/2-$
132	5.2657E-24	1.2500E+02	$N(1650)1/2-$

133	4.5394E-24	1.4500E+02	N(1675)5/2−
134	5.4851E-24	1.2000E+02	N(1680)5/2+
135	3.2911E-24	1.5000E+02	N(1700)3/2−
136	4.7105E-24	1.4000E+02	N(1710)1/2+
137	2.6328E-24	2.5000E+02	N(1720)3/2+
138	2.9918E-24	2.2000E+02	N(1875)3/2−
139	3.2910E-24	2.0000E+02	N(1900)3/2+
140	1.6455E-24	4.0000E+02	N(2190)7/2−
141	1.6455E-24	4.0000E+02	N(2220)9/2+
142	1.3164E-24	5.0000E+02	N(2250)7/2−
143	1.0126E-24	6.5000E+02	N(2600)11/2−

Δ baryon excitations

144	5.6257E-24	1.1700E+02	Δ(1232)3/2+
145	2.6328E-24	2.5000E+02	Δ(1600)3/2+
146	5.0361E-24	1.3000E+02	Δ(1620)1/2−
147	2.1940E-24	3.0000E+02	Δ(1700)3/2−
148	2.6328E-24	2.5000E+02	Δ(1900)1/2−
149	1.9946E-24	3.3000E+02	Δ(1905)5/2+
150	2.5187E-24	3.000E0+02	Δ(1910)1/2+
151	2.5187E-24	3.0000E+02	Δ(1920)3/2+
152	2.5178E-24	3.0000E+02	Δ(1930)5/2−
153	2.3095E-24	2.8500E+02	Δ(1950)7/2+
154	1.8806E-24	3.50000E+02	Δ(2200)7/2−
155	1.3164E-24	5.0000E+02	Δ(2420)11/2+

Λ hyperon excitations

156	1.3034E-23	5.0500E+01	Λ(1405)1/2−
157	4.2193E-23	1.5600E+01	Λ(1520)3/2−
158	4.3881E-24	1.5000E+02	Λ(1600)1/2+
159	1.8806E-23	3.5000E+01	Λ(1670)1/2−
160	1.0970E-23	6.0000E+01	Λ(1690)3/2−
161	2.1940E-24	3.0000E+02	Λ(1800)1/2−

162   4.3881E-24   1.5000E+02 $\Lambda(1810)1/2+$

163   8.2276E-24   8.0000E+01 $\Lambda(1820)5/2+$

164   6.9285E-24   9.5000E+01 $\Lambda(1830)5/2-$

165   6.5821E-24   1.0000E+02 $\Lambda(1890)3/2+$

166   3.2911E-24   2.0000E+02 $\Lambda(2100)7/2-$

167   3.2911E-24   2.0000E+02 $\Lambda(2110)5/2+$

168   4.3881E-24   1.5000E+02 $\Lambda(2350)9/2+$

Σ hyperon excitations

169   1.8284E-23   3.6000E+01 $\Sigma(1385)3/2+$

170   6.5821E-24   1.0000E+02 $\Sigma(1660)1/2+$

171   1.0970E-23   6.0000E+01 $\Sigma(1670)3/2-$

172   7.3135E-24   9.0000E+01 $\Sigma(1750)1/2-$

173   5.4851E-24   1.2000E+02 $\Sigma(1775)5/2-$

174   5.4851E-24   1.2000E+02 $\Sigma(1915)5/2+$

175   2.9919E-24   2.2000E+02 $\Sigma(1940)3/2-$

176   3.6567E-24   1.8000E+02 $\Sigma(2030)7/2+$

177   6.5821E-24   1.0000E+02 $\Sigma(2250)$

Ξ hyperon excitations

178   7.2331E-23   9.1000E+00 $\Xi(1530)°3/2+$

179   6.6486E-23   9.9000E+00 $\Xi(1530)^{-}3/2+$

180   2.7425E-23   2.4000E+01 $\Xi(1820)3/2-$

181   1.0970E-23   6.0000E+01 $\Xi(1950)$

182   3.2911E-23   2.0000E+01 $\Xi(2030)$

Ω hyperon excitation

183   1.1967E-23   5.5000E+01 $\Omega(2250)$

Λ_c charm hyperon excitations

184   2.5414E-22   2.5900E+00 $\Lambda_c(2595)^{+}1/2-$

185   9.7366E-23   6.7600E+01 $\Lambda_c(2860)^{+}3/2+$

186   1.1753E-22   5.6000E+00 $\Lambda_c(2880)^{+}5/2+$

187   3.2910E-23   2.0000E+01 $\Lambda_c(2940)^{+}3/2-$

Σ_c charm hyperon excitations

188	3.4826E-22	1.8900E+00	$\Sigma_c(2455)^{++}1/2+$
189	3.5968E-22	1.8300E+00	$\Sigma_c(2455)^o1/2+$
190	4.4534E-23	1.4780E+01	$\Sigma_c(2520)^{++}3/2+$
191	4.3020E-23	1.5300E+01	$\Sigma_c(2520)^o3/2+$
192	8.7762E-24	7.5000E+01	$\Sigma_c(2800)^{++}$
193	1.0616E-23	6.2000E+01	$\Sigma_c(2800)^+$
194	9.1418E-24	7.2000E+01	$\Sigma_c(2800)^o$

Ξ_c charm hyperon excitations

195	3.0758E-22	2.1400E+01	$\Xi_c(2645)^+$
196	2.8019E-22	2.3500E+01	$\Xi_c(2645)^0$
197	7.3956E-23	8.9000E+00	$\Xi_c(2790)^+$
198	6.5821E-23	1.0000E+01	$\Xi_c(2790)^0$
199	2.7097E-22	2.4300E+00	$\Xi_c(2815)^+$
200	2.5924E-22	2.5400E+00	$\Xi_c(2815)^0$
201	3.1494E-23	2.0900E+01	$\Xi_c(2970)^+$
202	2.3424E-23	2.8100E+01	$\Xi_c(2970)^o$
203	8.4386E-23	7.8000E+00	$\Xi_c(3055)^+$
204	1.8283E-22	3.6000E+00	$\Xi_c(3080)^+$
205	1.1754E-22	5.6000E+00	$\Xi_c(3080)^o$

Ω_c charm hyperon excitations

206	1.4627E-22	4.5000E+00	$\Omega_c(3000)^o$
207	7.5656E-23	8.7000E+00	$\Omega_c(3090)^o$

Σ_b bottom hyperon excitations

208	6.7856E-23	9.7000E+00	$\Sigma_b^+1/2+$
209	1.3433E-22	4.9000E+00	$\Sigma_b^-1/2+$
210	5.7325E-23	1.1500E+01	$\Sigma_b^{*+}3/2+$
211	8.7761E-23	7.5000E+00	$\Sigma_b^{*-}3/2+$

Ξ_b bottom hyperon excitations

212	7.3134E-22	9.0000E-01	$\Xi_b(5945)^03/2+$
213	3.9892E-22	1.6500E+00	$\Xi_b(5955)^-3/2+$

Appendix B. Particle Energy Database

This compilation is from M. Tanabashi *et al.* (Particle Data Group) *Phys. Rev.* D **98** 030001 (2018) (URL: http://pdg.lbl.gov).

Particle energies E and particle masses m are related by the Einstein equation $E = mc^2$. The particles shown here are the ones that have both experimentally-measured energies and lifetimes (see Appendix A). The isotopic-spin-averaged energies listed here are for comparison to the calculated isotopic-spin-averaged energy values in the text.

Energy (MeV)	Particle	Isotopic-spin-averaged energy
Long-lived threshold-state particles		
Leptons		
0.5109989461	$e^{\pm}$	
105.6583745	$\mu^{\pm}$	
1776.86	$\tau^{\pm}$	
Pseudoscalar mesons		
134.9770	π°	137.27 $\quad \pi$
139.57061	$\pi^{\pm j}$	

493.677   K$^\pm$ 495.66   K
497.611   K$^\circ$
547.862 η
957.78 η'

Baryons and hyperons

938.2720813   p   938.92   n
939.5654133   n
1115.683 Λ
1189.37 Σ^+
1192.642 Σ° 1193.15 Σ
1197.449 Σ^-
1314.86 Ξ° 1318.29 Ξ
1321.71 Ξ^-
1672.45 Ω^-

Charm mesons

1864.83   D$^\circ$ 1867.24   D
1869.65   D$^\pm$
1968.34   D$_s^\pm$
2010.26   D*(2010)$^\pm$
3096.900   J/Ψ_{1S}
3686.097 Ψ(P2s)

Charm hyperons

2286.46 Λ_c^+
2467.87 Ξ_c^+ 2469.4 $\in_c$
2470.87 Ξ_c°
2695.2 Ω_{cc}°

Bottom mesons

5279.32   B$^\pm$ 5279.5   B
5279.63   B$^\circ$

5366.89	B_s^o
6274.9	$B_c^\pm$
9460.30	Υ_{1S}
10023.26	Υ_{2S}
10355.2	Υ_{3S}

Bottom hyperons

5619.6	Λ_b
5794.5	Ξ_b^-
5791.9	Ξ_b^o
6046.1	Ω_b^-

Miscellaneous particles

782.65	$\omega(782)$
1019.461	$\phi(1020)$
2983.9	η_{c1S}
3637.6	η_{c2S}
80379.0	$W^\pm$
91187.6	Z^o
125180	Higgs
173000 ± 400	top quark

Short-lived excited-state particles

Unflavored mesons

775.26	$\rho^o(770)$
990.0	$f_o(980)$
980.0	$a_o(980)$
1170.0	$h_1(1170)$
1229.5	$b_1(1235)$
1275.5	$f_2(1270)$
1281.9	$f_1(1285)$

1294.0	$\eta(1295)$
1318.3	$a_2(1320)$
1354.0	$\pi_1(1400)$
1408.8	$\eta(1405)$
1426.4	$f_1(1420)$
1425.0	$\omega(1420)$
1474.0	$a_o(1450)$
1465.0	$\rho(1450)$
1476.0	$\eta(1475)$
1504.0	$f_o(1500)$
1525.0	$f'_2(1525)$
1662.0	$\pi_1(1600)$
1617.0	$\eta_2(1645)$
1670.0	$\omega(1650)$
1667.0	$\omega_3(1670)$
1672.2	$\pi_2(1670)$
1680.0	$\varphi(1680)$
1688.8	$\rho_3(1690)$
1720.0	$\rho(1700)$
1723.0	$f_o(1710)$
1812.0	$\pi(1800)$
1854.0	$\phi_3(1850)$
1895.0	$\pi_2(1880)$
1944.0	$f_2(1950)$
2011.0	$f_2(2010)$
1995.0	$a_4(2040)$
2018.0	$f_4(2050)$
2188.0	$\phi(2170)$
2297.0	$f_2(2300)$
2345.0	$f_2(2340)$

Strange mesons

891.76	$K*^{\pm}(892)$
895.55	$K*^{\circ}(892)$
1270.0	$K_1(1270)$
1403.0	$K_1(1400)$
1421.0	$K*(1410)$
1425.0	$K*_{\circ}(1430)$
1425.6	$K*_2(1430)^{\pm}$
1432.4	$K*_2(1430)^{\circ}$
1718.0	$K*(1680)$
1773.0	$K_2(1770)$
1776.0	$K*_2(1780)$
1819.0	$K_2(1820)$
2045.0	$K*_4(2045)$

Charm mesons

2318.0	$D^*_{\circ}(2400)^{\circ}$
2420.8	$D_1(2420)^{\circ}$
2460.7	$D^*_2(2460)^{\circ}$
2465.4	$D^*_2(2460)^{\pm}$

Bottom mesons

5726.0	$B_1(5721)^{\circ}$
5725.9	$B_1(5721)^{+}$
5737.2	$B^*_2(5747)^{+}$
5739.5	$B^*_2(5747)^{\circ}$
5851.0	$B_J(5840)^{+}$
5964.0	$B_J(5970)^{+}$
5871.0	$B_J(5840)^{\circ}$

Charm J/Ψ excitations

3414.71	$\eta_{c0}(1p)$
3510.67	$\eta_{c1}(1p)$

3525.38	$h(c)_{1P}$
3556.17	$\eta_{c2}(1p)$
3773.13	$\Psi(3770)$
3886.6	$Z_c(3900)$
3915.4	$X(3915)$
3927.2	$\eta_{c2}(3930)$
4024.1	$X(4020)$
4039.0	$\Psi(4040)$
4146.8	$\eta_{c1}(4140)$
4191.0	$\Psi(4160)$
4230.0	$\Psi(4260)$
4274.0	$\eta_{c1}(4274)$
4348.0	$\Psi(4360)$
4421.0	$\Psi(4415)$
4478.0	$Z_c(4430)$
4643.0	$\Psi(4660)$

Bottom Upsilon excitations

10579.4	Υ_{4s}
10889.9	$\Upsilon(10860)$
10992.9	$\Upsilon(11020)$

N baryon excitations

1440.0	$N(1440)1/2+$
1515.0	$N(1520)3/2+$
1530.0	$N(1535)1/2-$
1650.0	$N(1650)1/2-$
1680.0	$N(1675)5/2-$
1685.0	$N(1680)5/2+$
1720.0	$N(1700)3/2-$
1710.0	$N(1710)1/2+$
1720.0	$N(1720)3/2+$

1875.0	N(1875)3/2−
1880.0	N(1880)1/2+
1895.0	N(1895)1/2−
1920.0	N(1900)3/2+
2100.0	N(2060)5/2−
2100.0	N(2100)1/2+
2120.0	N(2120)3/2−
2180.0	N(2190)7/2−
2250.0	N(2220)9/2+
2280.0	N(2250)9/2−
2600.0	N(2600)11/2−

Δ baryon excitations

1232.0	Δ(1232)3/2+
1570.0	Δ(1600)3/2+
1610.0	Δ(1620)1/2−
1710.0	Δ(1700)3/2−
1860.0	Δ(1900)1/2−
1880.0	Δ(1905)5/2+
1900.0	Δ(1910)1/2+
1920.0	Δ(1920)3/2+
1950.0	Δ(1930)5/2−
1930.0	Δ(1950)7/2+
2200.0	Δ(2200)7/2−
2450.0	Δ(2420)11/2+

Λ hyperon excitations

1405.1	Λ(1405)1/2−
1519.5	Λ(1520)3/2−
1600.0	Λ(1600)1/2+
1670.0	Λ(1670)1/2−
1690.0	Λ(1690)3/2−
1800.0	Λ(1800)1/2−

1810.0	$\Lambda(1810)1/2+$
1820.0	$\Lambda(1820)5/2+$
1830.0	$\Lambda(1830)5/2-$
1890.0	$\Lambda(1890)3/2+$
2100.0	$\Lambda(2100)7/2-$
2110.0	$\Lambda(2110)5/2+$
2350.0	$\Lambda(2350)9/2+$

Σ hyperon excitations

1382.8	$\Sigma(1385)3/2+$
1660.0	$\Sigma(1660)1/2+$
1670.0	$\Sigma(1670)3/2-$
1750.0	$\Sigma(1750)1/2-$
1775.0	$\Sigma(1775)5/2-$
1915.0	$\Sigma(1915)5/2+$
1940.0	$\Sigma(1940)3/2-$
2030.0	$\Sigma(2030)7/2+$
2250.0	$\Sigma(2250)$

Ξ hyperon excitations

1531.80	$\Xi(1530)°3/2 + 1533.4 \ \Xi \ (1530)3/2+$
1535.0	$\Xi(1530)^-3/2+$
1823.0	$\Xi(1820)3/2-$
1950.0	$\Xi(1950)$
2025.0	$\Xi(2030)$

Ω hyperon excitation

| 2252.0 | $\Omega(2250)^-$ |

Λ_c charm hyperon excitations

2592.25	$\Lambda_c(2595)^+1/2-$
2856.1	$\Lambda_c(2860)^+3/2+$
2881.63	$\Lambda_c(2880)^+5/2+$
2939.6	$\Lambda_c(2940)^+3/2-$

Σ_c charm hyperon excitations

2453.97	$\Sigma_c(2455)^{++}1/2 + (2454)\Sigma_c(2455)1/2+$
2453.75	$\Sigma_c(2455)^o 1/2+$
2518.41	$\Sigma_c(2520)^{++}3/2 + (2518.45)\Sigma_c(2520)3/2+$
2518.48	$\Sigma_c(2520)^o 3/2+$
2801.0	$\Sigma_c(2800)^{++}$
2792.0	$\Sigma_c(2800)^+$
2806.0	$\Sigma_c(2800)^o$

Ξ_c charm hyperon excitations

2645.53	$\Xi_c(2645)^+$
2646.32	$\Xi_c(2645)^0$
2792.0	$\Xi_c(2790)^+$
2792.8	$\Xi_c(2790)^0$
2815.67	$\Xi_c(2815)^+$
2820.22	$\Xi_c(2815)^0$
2969.4	$\Xi_c(2970)^+$
2967.8	$\Xi_c(2970)^o$
3055.9	$\Xi_c(3055)^+$
3077.2	$\Xi_c(3080)^+$
3079.9	$\Xi_c(3080)^o$

Ω_c charm hyperon excitation

| 2765.9 | $\Omega_c(2770)^o$ |

(Note: $\Omega_c(2765.9)^o - \Omega_c^o(2695.2) = 70.7 \pm 0.9\,\text{MeV}$)*

| 3000.4 | $\Omega_c(3000)^o$ |
| 3090.2 | $\Omega_c(3090)^o$ |

Ξ_b bottom hyperon excitations

5811.3	$\Sigma_b^+ 1/2+$
5815.5	$\Sigma_b^- 1/2+$
5832.1	$\Sigma_b^{*+} 3/2+$
5836.1	$\Sigma_b^{*-} 3/2+$
5948.9	$\Xi_b(5945)^\circ 3/2+$
5955.33	$\Xi_b(5955)^- 3/2+$

Editor's comment: This note by Dr. Malcolm Mac Gregor draws attention to the famous mass quantum 70 MeV.

Appendix C. Magnetic Moment Values of u, d, s Constituent-Quark Energies

An independent determination of experimental fermion u, d, and s constituent-quark energies can be obtained from the measured hyperon magnetic moments [6]. The Dirac magnetic moment of a quark is $\mu_\mathrm{D} = q\hbar/2m_\mathrm{quark}c = q\hbar c/2E_\mathrm{quark}$, where q is the fractional charge on the quark and E_quark is the quark energy. The simplest quark model fit to the hyperon magnetic moments [6] gives the following quark energies: $E_u = 338\,\mathrm{MeV}$; $E_d = 322\,\mathrm{MeV}$; $E_s = 510\,\mathrm{MeV}$. The average of the E_u and E_d quark energies is $330\,\mathrm{MeV}$, which agrees with the calculated $3E_f = 315\,\mathrm{MeV}$ energy to 5% accuracy. The s quark energy is $E_s = 510\,\mathrm{MeV}$, which agrees with the calculated $5E_f = 525\,\mathrm{MeV}$ energy to 3% accuracy. The principal conclusion to be drawn from these magnetic moment results is not their numerical accuracy, but rather the fact that they establish *constituent-quark* energies [1] as the quantities that actually dictate quark and particle inertial-mass/energy properties.

[1] Malcolm H. Mac Gregor, *The Power of Alpha* (World Scientific, New Jersey, 2007), p. 57.

Index

www.ingramcontent.com/pod-product-compliance
Lightning Source LLC
Chambersburg PA
CBHW050631190326
41458CB00008B/2220